Contents

Preface

Engineering Materials: volume 1 meets the need for a comprehensive text on engineering materials. It introduces the basic science of materials technology including plain carbon steels, cast irons, heat treatment, non-ferrous metals and alloys, bearing metals, polymeric materials, the shaping of materials, and material testing. Thus it satisfies the requirements of the Business and Technician Education Council (BTEC) standard units U84/265, U84/266, U84/267, U84/268 and U84/272, and the text is eminently suitable for students studying for an engineering qualification at the ONC/OND level ('N' level).

This text leads naturally into the more specialised units of *Engineering Materials: volume 2*. Volume 2 is intended for students studying at the HNC/HND level and beyond.

The broad coverage of *Engineering Materials: volumes 1 & 2* ensures that they not only satisfy the requirements of Technician Engineers, but also provide an excellent technical background for undergraduates studying for a degree in Mechanical Engineering, Manufacturing Engineering, Materials Engineering or Combined Engineering.

The Author wishes to thank Mr. Tony May for reading the manuscript and providing many helpful comments and suggestions.

R. L. Timings
1991

Acknowledgements

We are indebted to the following for permission to reproduce copyright material:

GKN Bound Brook Ltd, for our Fig. 9.18 taken from *Porous Metal Bearings* by V.T. Morgan; Vandervell Ltd, for our Fig. 9.19.

R.L. Timings

En

vo

Longman Scientific & Technical,
Longman Group UK Limited,
Longman House, Burnt Mill, Harlow,
Essex CM20 2JE, England
and Associated Companies throughout the world.

First published 1989
Second (revised) impression 1991
Third impression 1991

British Library Cataloguing in Publication Data

Timings, R.L. (Roger Leslie), *1927-*
  Engineering materials.
  Vol. 1
  1. Materials science
  I. Title
  620.1′1

ISBN 0-582-42444-5

Set in Compugraphic Times 10/11 pt

Printed in Malaysia
by Chee Leong Press Sdn. Bhd.,
Ipoh, Perak Darul Ridzuan

1 Introduction to materials

1.1 Selection of materials

Figure 1.1 shows three objects. The first is a connector joining electric cables. The plastic casing has been partly cut away to show the metal connector. Plastic is used for the outer casing because it is a good electrical insulator and prevents electric shock if a person touches it. It also prevents the conductors touching each other and causing a short circuit. As well as being a good insulator the plastic is cheap, tough, and easily moulded to shape. It has been chosen for the casing because of these *properties*, that is, the properties of toughness, good electrical insulation, and ease of moulding to shape. It is also a relatively low cost material.

The metal conductor and its clamping screws are made from brass. This metal has been chosen because of its special properties. These properties are: good electrical conductivity, ease of machining to shape, adequate strength, and corrosion resistance. The precious metal silver is an even better conductor, but it would be far too expensive for this application and it would also be too weak and soft.

The second object in Fig. 1.1 is the connecting rod of a motor car engine. This is made from a special steel alloy. This alloy has been chosen because it combines the properties of strength, toughness, and the ability to be readily forged and machined to shape.

The third object shown in Fig. 1.1 is part of a machine tool. It is the tailstock casting for a small lathe. The metal used in this example is cast iron. This metal has been chosen because it combines the properties of adequate strength with ease of melting and casting to complicated shapes

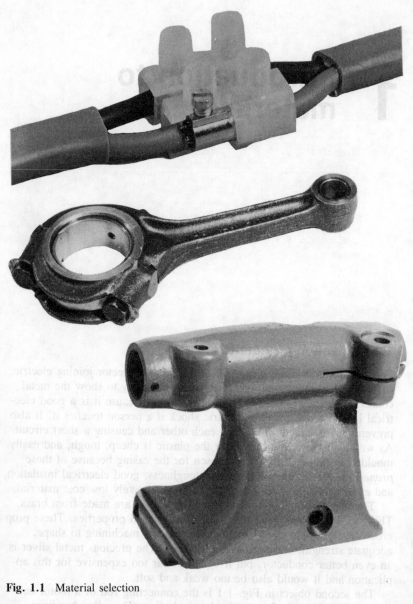

Fig. 1.1 Material selection

in simple sand moulds. It is also relatively easy to machine to its finished size and shape.

Thus the reasons for selecting the materials in the above example can be summarised as:

(*a*) Commercial factors such as:

(i)   cost,
(ii) availability,
(iii) ease of manufacture.
(b) Properties of materials such as:
(i)   electrical conductivity,
(ii) toughness,
(iii) corrosion resistance.

1.2   Properties of materials

The principal properties of materials which are of importance to the
engineer are as follows.

Tensile strength

This is the ability of a material to withstand tensile (stretching) loads
without breaking. Figure 1.2 shows a heavy load being held up by a rod
fastened to a beam. The load is trying to *stretch* the rod. Therefore the
rod is said to be in *tension*, so the material from which the rod is made
needs to have sufficient *tensile strength* to resist the pull of the load.

Compressive strength

This is the ability of a material to withstand compressive (squeezing)
loads without being crushed or broken. Figure 1.3 shows a component
being *compressed* by a heavy load. Therefore the component needs to be
made from a material with adequate *compressive strength* to resist the
load.

Shear strength

This is the ability of a material to withstand offset loads, or transverse
cutting (shearing actions). Figure 1.4(*a*) shows a rivet joining two metal
bars together. The forces acting on the two bars are trying to separate

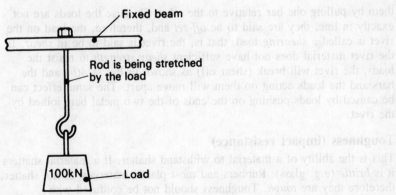

Fig. 1.2 Tensile strength

4

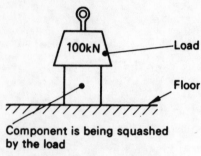

Component is being squashed by the load

Fig. 1.3 Compressive strength

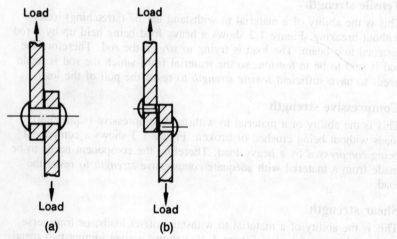

Fig. 1.4 Shear strength

them by pulling one bar relative to the other. Because the loads are not exactly in line, they are said to be *off-set* and, therefore, the load on the rivet is called a *shearing* load, that is, the rivet is said to be in *shear*. If the rivet material does not have sufficient *shear strength* to resist the loads, the rivet will break (shear off) as shown in Fig. 1.4(*b*) and the bars and the loads acting on them will move apart. The same effect can be caused by loads pushing on the ends of the two metal bars joined by the rivet.

Toughness (impact resistance)

This is the ability of a material to withstand shatter. If a material shatters it is *brittle* (e.g. glass). Rubbers and most plastic materials do not shatter, therefore they are *tough*. Toughness should not be confused with strength. Figure 1.5 shows a metal rod in a vice being subjected to

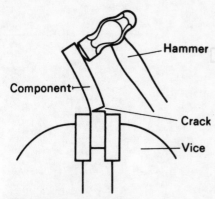

Fig. 1.5 Impact strength

impact loading. If the rod is made from a piece of high carbon steel, for example silver steel in the annealed (soft) condition as supplied, it will have only a moderate tensile strength, but under the impact of the hammer it will bend without breaking, therefore it is *tough*. If a similar specimen is made hard by making it red-hot and cooling it quickly in water, it will now have a very much higher tensile strength. However, although it is now *stronger* it will prove to be brittle and will break off easily when struck with a hammer. Therefore it now *lacks toughness*. A material is brittle and shatters because small cracks in its surface grow quickly under a tensile force. (As a material bends its outer layers tend to stretch and are in tension.) Therefore any material in which the spread of surface cracks does not occur or only occurs to a small extent is said to be *tough*.

Elasticity

This is the ability of a material to deform under load and return to its original size and shape when the load is removed. Figure 1.6 shows a tensile test specimen. If it is made from an elastic material it will be the same length before and after the load is applied, despite the fact that it will be longer whilst the load is being applied. This is only true for most materials, if the load is relatively small and within the elastic range of the material being tested.

Plasticity

This property is the exact opposite to elasticity. It is the state of a material which has been loaded beyond the elastic state. Under a load beyond that required to cause elastic deformation the material deforms permanently. It takes a *permanent* set and will not return to its original size and shape when the load is removed. When a piece of mild steel strip is bent at right-angles into the shape of a bracket, it shows the property of plasticity since it does not spring back straight again. This is

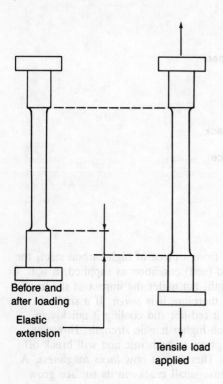

Before and after loading

Elastic extension

Tensile load applied

Fig. 1.6 Elasticity

shown in Fig. 1.7. *Ductility* and *malleability* are particular cases of the property of plasticity and they will now be considered separately.

Ductility

This is the term used when plastic deformation occurs as the result of applying a *tensile* load. A ductile material is required for such processes as wire drawing (Fig. 1.8), tube drawing, and cold-pressing low-carbon steel sheets into motor car body panels.

Malleability

This is the term used when plastic deformation occurs as the result of applying a *compressive* load. A malleable material is required for such processes as forging, rolling, and rivet heading as shown in Fig. 1.9.

Hardness

This is defined as the ability of a material to withstand scratching (abrasion) or indentation by another hard body. It is an indication of the wear resistance of the material. Figure 1.10 shows a hardened steel ball being

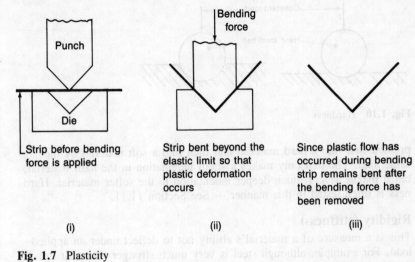

Fig. 1.7 Plasticity

(i)   (ii)   (iii)

Strip before bending force is applied

Strip bent beyond the elastic limit so that plastic deformation occurs

Since plastic flow has occurred during bending strip remains bent after the bending force has been removed

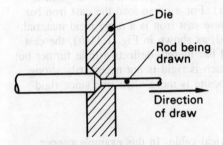

A rod being drawn through a die to reduce its diameter requires the property of ductility

Fig. 1.8 Ductility

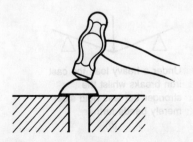

Forming the head of a rivet by hammering. The rivet needs to be made from a malleable material to withstand this treatment

Fig. 1.9 Malleability

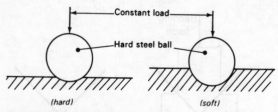

Fig. 1.10 Hardness

pressed first into a hard material and then into a soft material by the same load. The ball only makes a small indentation in the hard material, but it makes a very much deeper indentation in the softer material. Hardness is often tested in this manner — See Section 11.11.

Rigidity (stiffness)

This is a measure of a material's ability not to deflect under an applied load. For example, although steel is very much stronger than cast iron the latter material is preferred for machine beds and frames because it is more rigid and less likely to deflect with consequent loss of alignment and accuracy. Consider Fig. 1.11(a). For a given load the cast iron bar deflects less than the steel bar because cast iron is a more rigid material. However when the load is increased, as shown in Fig. 1.11(b), the cast iron bar will break, whilst the steel bar merely deflects a little further but does not break. Thus a material which is rigid is not necessarily strong. In fact, as has been shown, the opposite is more often true since rigid materials are often brittle.

Electrical conductivity

Figure 1.12 shows a piece of electrical cable. In this example copper wire has been chosen for the conductor or core of the cable because cop-

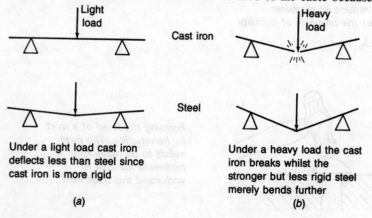

Fig. 1.11 Rigidity

per has the property of very good *electrical conductivity*, that is, it offers very little resistance to the flow of electrons (electric current) through the wire. A plastic material such as polyvinyl chloride (PVC) has been chosen for the insulating sheathing surrounding the wire conductor. This material has been chosen because it is such a bad conductor that very few electrons can pass through it. Very bad conductors such as PVC are called *insulators*. There is no such thing as a perfect insulator, only very bad conductors. Care must be taken when comparing and interpreting tables of test data. For example, metallic conductors of electricity all increase in resistance as their temperatures rise. Pure metals show this effect more strongly than alloys. However, pure metals generally have a better conductivity than alloys at room temperature. The conductivity of metals and metal alloys improves as the temperature falls. Conversely, non-metallic materials used for insulators tend to offer a lower resistance to the passage of electrons, and so become poorer insulators, as their temperatures rise. Thus care must be taken if cables are bunched together without adequate ventilation. Glass, for example, is an excellent insulator at room temperature, but becomes a conductor if raised to red-heat. Table 1.1 lists the properties of typical conductor materials, and Table 1.2 lists the properties of typical insulating materials.

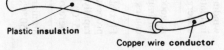

Plastic insulation

Copper wire conductor

Fig. 1.12 Electrical conductivity

Table 1.1 Properties of conductor materials

Material	Resistivity ($\mu\Omega$ mm)	Temperature coefficient α_0 (per °C)
Aluminium	28.0	42×10^{-4}
Carbon (graphitic) (1)	46.0×10^3	-5×10^{-4}
Copper (annealed)	17.2	43×10^{-4}
Mild steel	107.0	65×10^{-4}
Manganin alloy (2)	480.0	0
Nichrome alloy (3)	1090.0	53×10^{-4}
Nickel	136.0	56×10^{-4}
Silver (annealed)	15.8	41×10^{-4}

Note: (1) The negative sign in the temperature coefficient column indicates a fall in resistance as temperature increases.

(2) *Manganin* is an alloy of copper, manganese and nickel and is used for wire wound resistors for use in measuring instruments where its zero value of temperature coefficient means that the ohmic value of such resistors are unaffected by temperature change.

(3) *Nichrome* is an alloy of nickel and chromium which resists oxidation at high temperatures. It is used for heating elements in electric radiators and furnaces.

Table 1.2 Properties of insulating materials

Material	Typical applications	Properties
Insulating oil	Switch gear, transformers, cable impregnation	Reduces arcing between switchgear contacts. Being a fluid insulator it can circulate by convection and cool the windings and cores of large transformers as well as insulating them. Used to impregnate the paper insulation in underground armoured mains cables.
Paper	Armoured mains cables	A good, relatively cheap, insulator for rigid cables. Must be oil impregnated and sealed against ingress of moisture which causes the oil to break down.
Rubber (natural)	Flexible cables	Flexible, high insulation resistance, reasonable mechanical properties when 'vulcanised' with 5% sulphur. Degrades rapidly (perishes) in strong sunlight and undue heat (max 55°C). Sulphur content attacks copper conductors which, therefore, must be tinned.
Silicone rubber (synthetic)	Flexible cables	Similar to rubber, but suitable for applications up to 150°C, very much more expensive.

Polyvinyl chloride (PVC)	Flexible cables	Although PVC has a much lower insulation resistance than rubber it is now more widely used because of its low cost and resistance to oils, petrol and chemical solvents. Max operating temperature 65°C.
Mineral insulation (magnesium oxide)	Rigid metal sheathed (MIMS) cables sheathed heating elements	Can operate at temperatures up to dull-red heat. Limited to 660 Volts. Terminations must be sealed as magnesium oxide powder is hygroscopic.
Thermosetting plastics	Moulded insulators (interior)	(i) Phenolic resins (Bakelite) used for moulded insulation blocks and switchgear components where strength is important and its dark colour acceptable. (ii) Amino resins are used for domestic switchgear mouldings as it is available in white and light colours. Lower strength than phenolic resin.
Glass & ceramics	Moulded insulators (exterior)	Hard, highly glazed surface, prevents weathering. Weak in tension and shear, insulators must be designed to operate under compressive mechanical loads only. Used for high voltage insulators for overhead transmission lines. High insulation resistance. Woven glass fibre (resin varnish impregnated) used for high temperature insulation and sheathing in domestic cookers.

Magnetic properties

Just as some materials are good or bad conductors of electricity, some materials can be good or bad conductors of magnetism. The good magnetic conductors are the *ferro-magnetic* materials which get their name from the fact that they are made from iron, steel and associated alloying elements such as cobalt and nickel. All other materials are non-magnetic and offer a high *reluctance* (resistance) to the magnetic flux field. Magnetic materials can be classified as follows.

(a) *Hard magnetic materials.* These retain their magnetism after the initial magnetising force has been removed. Traditionally, permanent magnets were made from quench-hardened high-carbon steels. Modern permanent magnet alloys are more powerful than high carbon steels and are used for all but the most simple low-cost applications. These materials get their name from the fact that they are very hard and cannot be machined except by grinding. Table 1.3 lists some typical ferro-magnetic alloys suitable for permanent magnets together with their corresponding BH_{max} values. The BH_{max} value is the maximum magnetic energy which the magnet can give out. It can be seen from the table that an alloy such as 'Columax' is nearly 34 times more powerful than a similar magnet made from high-carbon steel. However, although these materials retain their magnetism very well once they have been magnetised, they have a *low magnetic permeability* and this makes them difficult to magnetise initially.

(b) *Soft magnetic materials.* These show properties which are the opposite of those just described for hard magnetic materials. Soft magnetic materials have *high magnetic permeability*. This means that they are very easily magnetised and demagnetised and that they retain virtually no magnetism when the magnetising force is removed. Traditionally, low carbon steels and wrought iron were used for transformer, choke, and electromagnet cores and for generator and motor pole pieces. Nowadays, however, these materials have been superseded by special alloys for such applications and these alloys possess very low hysteresis values and very high magnetic permeability. Two such alloys will now be described.

(i) *Mumetal.* This is an alloy containing 74 per cent nickel, 5 per cent copper, 1 per cent manganese, and 20 per cent iron. This is an expensive alloy, but is widely used as a shielding material in telecommunications equipment.

(ii) *Silicon iron.* This contains 4.0 per cent silicon, 0.3 per cent manganese, less than 0.05 per cent carbon, and the remainder iron. This alloy is very much cheaper than 'Mumetal' whilst still possessing very low hysteresis values and high magnetic permeability. It is widely used for lamination stampings for the cores for transformers, chokes, and motor and generator rotors and stators.

Thermal conductivity

This is the ability of a material to transmit heat energy by conduction. Figure 1.13 shows a soldering iron. The bit is made from copper which is a good conductor of heat and so will allow the heat energy stored in it to travel easily down to the tip and into the work being soldered. The wooden handle remains cool as it has a low thermal conductivity and resists the flow of heat energy. Table 1.4 lists the thermal properties of a number of engineering materials.

Fusibility

This is the ease with which materials will melt. It can be seen from Fig. 1.14 that solder melts easily and so has the property of *high fusibility*. On the other hand, fire bricks used for furnace linings only melt at very high temperatures and so have the property of *low fusibility*. Such materials which only melt at very high temperatures are called *refractory materials*. These must not be confused with materials which have a low thermal conductivity and are used as thermal insulators. Reference to Table 1.4 shows that although expanded polystyrene is an excellent thermal insulator, it has a very low melting point (high fusibility) and in no way can it be considered a refractory material.

Temperature stability

Substantial changes in temperature can have very significant effects on the structure and properties of materials and these will be considered later. However, there are two effects which changes in room temperature can have on the dimensional stability of a component.

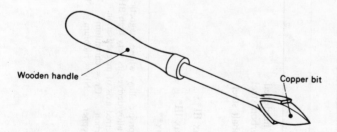

Wooden handle

Copper bit

Fig. 1.13 Thermal conductivity

Table 1.3 Permanent magnet materials

Name	Composition (%)										BH_{max}
	C	Cr	W	Co	Al	Ni	Cu	Nb	Ti	Fe	
Quench hardened High-carbon steel	1.0	—	—	—	—	—	—	—	—		1 560
35% cobalt steel	0.9	6.0	5.0	35.0	—	—	—	—	—		7 800
Alnico	—	—	—	12.0	9.5	17.0	5.0	—	—	Remainder	13 500
Alcomax III*	—	—	—	24.5	8.0	13.5	3.0	0.6	—	←	38 000
Hycomax III*	—	—	—	34.0	7.0	15.0	4.0	—	5.0	Remainder	35 200
Columax**	—	—	—	24.5	8.0	13.5	3.0	0.6	—	→	52 800

* Anisotropic alloys whose magnetic properties are measured along the preferred axis
** This alloy derives its very high BH value from the way it is cooled during casting which orientates its columnar crystals parallel to the preferred axis of magnetisation

C = carbon   Cr = chromium   W = tungsten   Al = aluminium   Ni = nickel
Cu = copper   Nb = niobium   Ti = titanium   Co = cobalt   Fe = iron

Table 1.4 Thermal properties of materials

Material	Melting point	Conductivity
Aluminium	660°C	Very good
Copper	1080°C	Excellent
Iron	1535°C	Good
Wood	No melting point (burns)	Poor (good insulator)
Polystyrene (expanded)	No defined melting point but softens at 100°C	Very poor (Excellent insulator)
Glass Fibre	No defined melting point but softens at about 600°C to 800°C depending on composition	Very poor (very good insulator)
Fire bricks & clays for furnace linings	1595°C to 1800°C depending upon alumina content softening and loss of strength is progressive and commences below these temperatures	Poor (good insulator)

(a) *Thermal expansion.* All materials, to a greater or lesser extent, expand when heated and contract when cooled. This expansion or contraction is proportional to the change in temperature.

(b) *Creep.* This is an important factor when considering polymeric (plastic) materials. It must also be considered when metals work continuously at high temperatures, for example gas turbine blades. Creep is defined as a gradual extension of a material over a long period of time while under applied load or kept constant. The creep rate increases if the temperature is raised, but becomes less if the temperature is reduced. Creep will be considered in greater detail in section 1.2.

1.3 Engineering materials

Almost every type of material that exists has found its way into the engineering workshop at one time or another. The most convenient way to study the properties and uses of engineering materials is to group them into 'families'.

Ferrous metals

These are metals and alloys which contain a high proportion of the element iron. They are the most important and widely used in applications where high strength is required at relatively low cost and where weight is not of primary importance, for example, bridge building, the structure of large buildings, railway lines, locomotives and rolling stock and the bodies and highly stressed engine parts of road vehicles. However, light weight alloys such as aluminium and synthetic polymers (plastics) are being increasingly used in railway and road vehicles to reduce their weight over the

Non-ferrous metals

These materials refer to all the remaining metals known to man. They are rarely used as structural materials with the exception of pure copper and

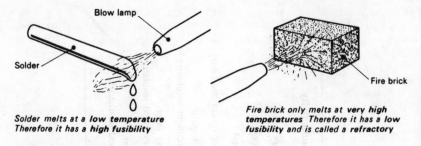

Fig. 1.14 Fusibility

(a) *Thermal expansion.* All materials, to a greater or lesser extent, expand when heated and contract when cooled. This expansion or contraction is proportional to the change in temperature.

(b) *Creep.* This is an important factor when considering polymeric (plastic) materials. It must also be considered when metals work continuously at high temperatures, for example gas-turbine blades. Creep is defined as the gradual extension of a material over a long period of time whilst the applied load is kept constant. The creep rate increases if the temperature is raised, but becomes less if the temperature is lowered. Creep will be considered in greater detail in Section 13.2.

1.3 Engineering materials

Almost every substance known to man has found its way into the engineering workshop at some time or other. The most convenient way to study the properties and uses of engineering materials is to group them into 'families'.

Ferrous metals

These are metals and alloys containing a high proportion of the element *iron*. They are the strongest materials available and are used for applications where high strength is required at relatively low cost and where weight is not of primary importance, for example, bridge building, the structure of large buildings, railway lines, locomotives and rolling stock and the bodies and highly stressed engine parts of road vehicles. However, light weight materials such as aluminium alloys and polymers (plastics) are being increasingly used in railway and road vehicles to reduce their weight and make them more efficient in the use of energy.

Non-ferrous metals

These materials refer to all the remaining metals known to man. They are rarely used as structural materials with the exception of pure copper and aluminium and these are mainly used where such properties as corrosion resistance, electrical conductivity and thermal conductivity are required. Copper is much stronger than aluminium, but also much heavier. However copper is much easier to join by such processes as soldering and brazing.

Non-ferrous alloys

(a) *Copper alloys.* Alloys such as bronze are highly corrosion resistant, strong, easily machined and have a relatively high melting point. Unfortunately they are heavier and much more costly than ferrous metals and alloys. Bronze alloys are widely used for steam and hydraulic valve components and for marine applications. The brass alloys are cheaper than the bronzes but are weaker and rather less corrosion resistant. Brass alloys are easily flow formed to shape and easily machined to a good finish. They are widely used in the manufacture of electrical components and domestic water fittings. These alloys are considered in greater detail in Chapter 7.

(b) *Aluminium alloys.* These materials are generally less strong than the ferrous- and copper-based alloys but have the advantage of being much lighter in weight. Like the copper alloys they are much more corrosion resistant than the ferrous metals and alloys, with the exception of the 'stainless steels'. Aluminium alloys are widely used in aircraft construction where low weight is of paramount importance. Unfortunately the strength of these alloys falls off rapidly as their temperature rises and for this reason supersonic aircraft are made from titanium alloy. Titanium is also a non-ferrous metal as light in weight as aluminium. However it is as strong as steel and retains its strength at high temperatures. Unfortunately it is very costly and relatively difficult to shape. Non-ferrous alloys are considered in greater detail in Chapter 7.

Non-metals (synthetic)

These so-called 'plastic' materials are used for an ever increasing range of duties and, because they are synthetic, new materials are being created all the time. Generally they combine good corrosion resistance with ease of manufacture by moulding to shape and relatively low cost. Their properties can be matched to almost any design problem with the exception of high temperature environments. Some of these materials can match the strength of alloy steels on a weight for weight basis, and are now replacing metals for many traditional applications. Synthetic adhesives are also being used for joining metallic components even in highly stressed applications. These materials are considered in greater detail in Chapter 8.

Non-metals (natural)

Such materials are so diverse that only a few can be listed here to give a basic introduction to some typical applications.

Wood	This is used for the manufacture of casting patterns.
Rubber	This is used for hydraulic and compressed air hoses and oil seals.
Glass	This is used for optical components in measuring instruments and, in the form of fibres, is used to reinforce plastics.
Emery	This is a widely used abrasive and is a naturally occurring aluminium oxide. Nowadays it is produced synthetically to maintain uniform quality and performance.
Ceramics	High voltage insulators and high temperature resistant cutting tool tips.
Diamonds	Cutting tools for operation at high speeds for metal finishing where surface finish is of great importance, for example, internal combustion engine pistons and bearings.
Oils	Bearing lubricants. Cutting fluids. Fuels.

1.4 Factors affecting material properties

The following are the more important factors which can influence the
properties and performance of engineering materials.

Heat treatment

This is the controlled heating and cooling of metals to change their properties to improve their performance or to facilitate processing. Heat treatment processes are considered in detail in Chapters 5 and 7. An example
of heat treatment is the hardening of a piece of high carbon steel rod. If
it is heated to a dull red heat and plunged into cold water to cool it
rapidly (quenching), it will become hard and brittle. If it is again heated
to dull red heat but allowed to cool slowly it will become softer and less
brittle (more tough). In this condition it is said to be *annealed*. When annealed it will be too soft and the grain will be too coarse for it to
machine to a good surface finish, but it will be in its best condition for
flow forming. During forming the grains will be distorted and this will
result in the metal becoming *work-hardened* if it is flow formed at room
temperature. To remove any locked-in stresses from the forming operations and to prepare the material for machining, the material has to be
normalised. This consists of again heating the metal to dull red heat but
cooling slowly in still air away from draughts. This results in a finer
grain than annealing and improves the metal's machining properties and
strength. Thus by various heat treatment processes a given material can
be made to exhibit a range of properties.

Processing

Hot and cold working processes will be referred to many times
throughout this book and it is essential to understand what is meant by
the terms *hot* and *cold working* as applied to metals. Basically, metals are
said to be *worked* when they are squeezed or stretched into shape. Metals

which have been squeezed or stretched to shape are said to be in the *wrought* condition. Thus metal which is worked into shape must possess the property of plasticity as described earlier in this chapter. Figure 1.15 shows examples of hot and cold working.

Metal is *hot-worked* or *cold-worked* depending upon the temperature at which it is flow formed to shape. These temperatures are not easy to define. For instance, lead hot-works at room temperature and can be beaten into complex shapes without cracking, but steel does not hot-work until it is red hot.

When metals are examined under the microscope it can be seen that they consist of very small crystals or grains. When most metals are bent or worked at room temperature these crystals become distorted and the metal becomes hard and brittle. That is why metal which has been *cold-worked* becomes *work-hardened*. Care must be taken to avoid excessive cold-working as this could cause the metal to crack. If considerable working is required to form a particular component, the metal must be softened from time to time during the processing by heat treatment (annealing). When metals are *hot-worked* the crystals are also distorted. However they reform instantly into normal crystals because the process temperature is above the temperature of *recrystallisation* for the metal being used and work-hardening does not occur. Thus *cold-working* is the flow forming of metals, *below* the temperature of recrystallisation, whilst *hot-working* is the flow forming of metals *above* the temperature of recrystallisation. Hot- and cold-working together with recrystallisation will be considered in greater detail in Section 5.1.

Environmental reactions

The properties of materials can also be affected by reaction with the environment in which they are used. For example:

(a) *De-zincification of brass*. Brass is an alloy of copper and zinc and when brass is exposed to a marine environment for a long time, the

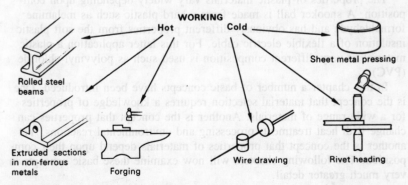

Fig. 1.15 Example of hot- and cold-working

salts in the sea water spray react with the zinc content of the brass so as to remove it and leave behind a spongy, porous mass of copper. This obviously weakens the component which fails under normal working conditions.

(b) *Rusting of steel*. Unless steel structures are regularly maintained by rust neutralisation and painting processes, rusting will occur. The rust will eat into the steel, reduce its thickness and, therefore, its strength, and weaken the structure. In extreme cases the entire structure may be eaten away.

(c) *Degradation of plastics*. Many plastics degrade and become weak and brittle when exposed to the ultra-violet content of sun-light and special dye-stuffs have to be incorporated into the plastic to filter out these harmful rays. This will be considered further in Section 13.13.

1.5 Composition

The properties of a material depend largely upon the composition of the material. For example, consider the plain carbon steels. These are alloys of iron and carbon. If the carbon content is under 0.3 per cent, then the alloy is referred to as low-carbon steel. This is relatively soft and ductile with moderate strength and it cannot be hardened by heating and quenching. However, if the carbon content is increased to 1.2 per cent the alloy is referred to as high-carbon steel. This is much less ductile, but is stronger and it can be hardened by heating and cooling rapidly (quenching) to make cutting tools.

Again, a brass alloy containing 70 per cent copper and 30 per cent zinc has very great ductility and can be cold-worked into complicated shapes without cracking. On the other hand, a brass alloy containing 60 per cent copper and 40 per cent zinc lacks ductility and cannot be readily cold-worked. However, it hot-works excellently and is widely used for hot-stamping into such shapes as plumbing fittings, water taps, etc. It is widely used for hot extrusion into rods and sections.

The properties of plastic materials vary widely depending upon composition. A snooker ball is made from a hard plastic such as melamine formaldehyde and has obviously different properties from the soft plastic insulation of a flexible electric cable. For this latter application a plastic material of very different composition is used such as polyvinyl chloride (PVC).

In this chapter a number of basic concepts have been introduced. One is the concept that material selection requires a knowledge of properties for a wide range of materials. Another is the concept that properties can change with heat treatment, processing and environmental reactions. Yet another is the concept that properties of materials depend upon their composition. The following chapters will now examine these basic concepts in very much greater detail.

2 Basic science of materials

2.1 Introduction

Figure 1.1 showed a number of components made from metal. Although the properties of the metals used varied widely they all had one thing in common. No matter what their composition, no matter what changes they had undergone during extraction from the ore, refinement and processing, they were all *crystalline*. However, before studying the crystalline structure of metals and alloys it is advisable to revise some basic concepts.

2.2 Atoms

Atoms can be considered as the smallest particles of substance which can exhibit all the properties of that substance. An atom is electrically neutral because it has an equal number of negatively charged electrons and positively charged protons. Figure 2.1(a) shows, diagramatically, an atom of the gas hydrogen (the simplest atom) and Fig. 2.1(b) shows an atom of the metal copper. It can be seen that both atoms consist of a *nucleus* around which it can be considered that there are orbiting one or more *electrons*. Although the electrons spend most of their time in 'shells' as shown in Fig. 2.1(b), these are not as rigid as the figure implies and it is now assumed that the electrons are free to move around the nucleus to form a 'cloud' as shown in Fig. 2.1(c). The basic structure of the atom is as follows.

Nucleus   This is the basic core of the atom and consists of *protons* and *neutrons*.

Protons   These are *positively* charged particles of very much greater mass than the electrons.

Neutrons   These particles have the same mass as protons but carry no electrical charge. The mass of the atom is generally considered to be the sum of the mass of the protons and neutrons

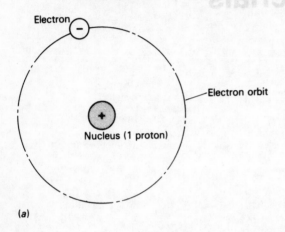

Electron

Electron orbit

Nucleus (1 proton)

(a)

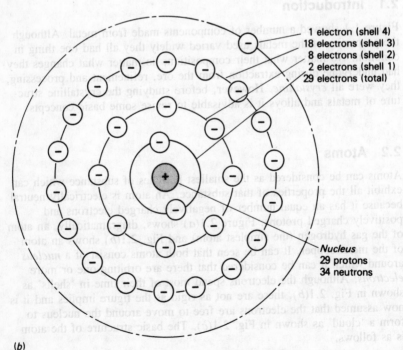

1 electron (shell 4)
18 electrons (shell 3)
8 electrons (shell 2)
2 electrons (shell 1)
29 electrons (total)

Nucleus
29 protons
34 neutrons

(b)

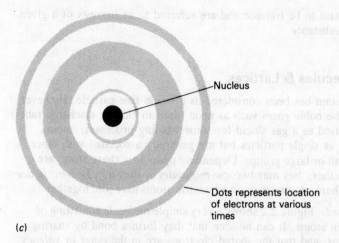

Fig. 2.1 Typical atomic structures (*a*) hydrogen atom, (*b*) copper atom, (*c*) movement of electrons round the nucleus

	present since the mass of the electrons is negligible by comparison.
Electrons	These particles are *negatively* charged and can be considered to orbit the nucleus like planets around the sun. Although electrons are very small and have only 1/1836 the mass of a proton or a neutron, they are supremely important when considering how atoms bond together to form molecules. The chemical properties of an atom, that is how it combines with other atoms, are determined by the number of electrons it has. Electrons also determine the electrical and magnetic properties of a material.
Ions	Ions are atoms which have gained or lost one or more electrons. Loss of an electron makes the atom *electropositive* since there will be a positively charged proton without its balancing electron. Such an ion is called a *positive ion*. Gaining an electron makes the atom *electronegative* since there is no spare positively charged proton in the nucleus to balance the additional electron. Such an ion is called a *negative ion*.
Isotopes	Since the electron is so small compared with the proton, the mass of the atom can — for all practical purposes — be considered as concentrated in the nucleus. As has already been stated, the neutron has the same mass as the proton but with no electrical charge. Thus, if the number of neutrons in the nucleus changes, the mass of the atom changes but its chemical properties do not change since there has been no change in the number of electrons present. Atoms which have the same chemical properties but different atomic masses are

said to be *isotopic* and are referred to as *isotopes* of a given substance.

2.3 Molecules & Lattices

So far, the atom has been considered as a single free particle. However, apart from the noble gases such as neon (used in electric discharge tubes) and argon (used as a gas shield for some welding processes), atoms rarely occur as single particles but are generally associated with other atoms in small or large groups. Depending upon how these atoms are grouped together, they may become *molecules* or they may become *lattice structures*. There are two ways in which the atoms may join together.

Covalent bond   Figure 2.2 shows a very simple *molecule* consisting of two hydrogen atoms. It can be seen that they form a bond by sharing their electrons, and that the shared electrons are in the outer or *valency* shell of the atom. Since the bond is formed by sharing electrons it is referred to as a *covalent bond*.

Ionic bond   This is formed by the complete transference of an electron from one atom to another. Compounds produced by ionic bounding do NOT form molecules but have a *lattice structure*, as do all other crystalline solids including the metals. For example, in common salt (sodium chloride), one isolated sodium atom has only one electron in its outer (valency) shell, but eight electrons in its next, inner shell. One isolated chlorine atom has seven electrons in its outer (valency) shell. In becoming a stable compound both atoms need to have 'complete' outer shells of

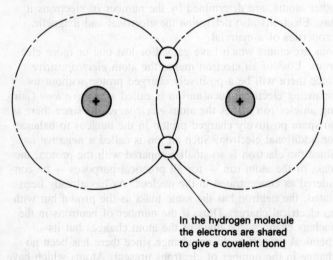

In the hydrogen molecule the electrons are shared to give a covalent bond

Fig. 2.2   A covalent bond

eight electrons. To achieve this, the sodium atom gives up the sole electron in its outer shell and becomes electro-positive. The chlorine atom gains this electron and becomes electro-negative, that is, both atoms become ions. Since particles with unlike charges attract each other, the sodium ion is attracted to the chlorine atom and an *ionic bond* is formed. At the same time both atoms attain full (eight electron) valency shells. This is shown in Fig. 2.3.

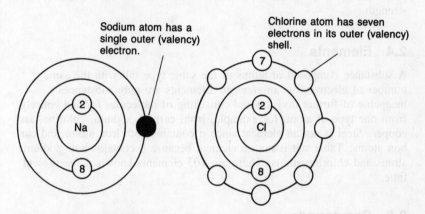

Sodium atom has a single outer (valency) electron.

Chlorine atom has seven electrons in its outer (valency) shell.

(a) Isolated atoms of sodium & chlorine

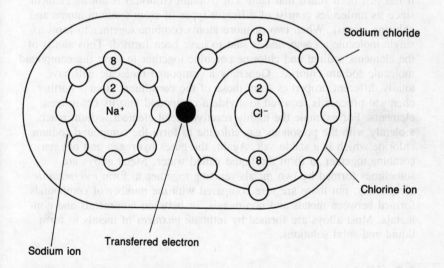

Sodium chloride

Chlorine ion

Sodium ion

Transferred electron

(b) Ionic bond formed by transfer of electron

Fig. 2.3 Ionic bond

Just as atoms are held together by powerful *primary bonds* to form molecules, so molecules can be bonded together by weaker *secondary* electrostatic forces. These forces are called *Van der Waal's forces* after the Dutch physicist Johannes Diderik Van der Waal. The secondary electrostatic forces bonding molecules together influence such properties as melting point, solubility, and the tensile strength of materials (particularly polymers). Adhesives of the 'impact' type rely on these secondary intermolecular forces. Although the attractive force between each pair of molecules is relatively weak, collectively they show great strength.

2.4 Elements

A substance composed of atoms of the same type (all with the same number of electrons) is an *element*. Elements are pure substances incapable of further division and consisting of molecules formed entirely from one type of atom, for example, iron, carbon, sodium, chlorine, and copper. Steel is not an element since it contains both iron atoms and carbon atoms. Table salt is not an element because it contains both sodium atoms and chlorine atoms. There are 103 elements known at the present time.

2.5 Compounds

It has just been stated that table salt (sodium chloride) is not an element since its molecules consist of different types of atom (sodium atoms and chlorine atoms). When two or more atoms combine together to form a single molecule a *compound* is said to have been formed. Thus atoms of the elements sodium and chlorine combine together to form the compound molecule sodium chloride. Generally a compound molecule will have totally different properties from those of the constituent atoms. Further, a chemical process is required to divide a compound into its constituent elements. For example the highly reactive metal element sodium reacts violently with the poisonous gas chlorine to form the compound sodium chloride which is a stable salt. Again, the gases hydrogen and oxygen combine together to form the liquid called water. Metal alloys are sometimes formed by two metals reacting together to form *intermetallic compounds*, but these are rare compared with the number of compounds formed between metals and non-metals, or between non-metals and non-metals. Most alloys are formed by intimate *mixtures* of metals to form liquid and solid solutions.

2.6 Mixtures

Mixtures are formed when two or more substances are in close association (mixed together) without any chemical reaction taking place, without any new substance (compound) being formed, and with the individual substances being separable by physical means. For example iron filings and sand can be mixed together. No new substance is formed and the iron filings can be removed with a magnet. Again, a mixture of sand and salt can be separated by dissolving the salt in water, filtering off the sand and recovering the salt by boiling the water so that evaporation occurs.

2.7 Solids, liquids and gases

Most substances can exist as solids, liquids or gases depending upon their temperature. A notable exception is iodine which *sublimes* directly from the solid state into the gaseous state when heated, without becoming a liquid. Most substances behave like water. Below its freezing point water is a solid (ice). Above its freezing point and below its boiling point it is in the liquid state. If its temperature is increased still further it boils and becomes a gas (steam). The fact that a substance can exist in the solid state, the liquid state, and the gaseous state, is due to the fact that the atoms and molecules of a substance are in a permanent state of vibration providing the temperature is above *absolute zero* ($-273°C$), at which temperature all movement stops. When the temperature is sufficiently low for a given substance to be in its solid state, the vibration is of small amplitude and the atoms and molecules only move to a small extent about a fixed point. When the temperature of a solid is raised to above its melting point, the atoms and molecules of the substance vibrate more violently and no longer move about a fixed position but are free to move about within the constraints of the container holding the liquid. Finally, if the temperature is raised still further until it is above the temperature of vaporisation for the substance, the atoms and molecules move so freely that they can disperse until they completely fill the vessel containing them and, if they escape, they continue to disperse throughout the atmosphere.

The changes of state of a substance do not proceed evenly as the temperature rises, but in a series of steps as shown in Fig. 2.4. This is due to a change in state being accompanied by the taking in or the giving out of *latent heat*, that is, the heat energy associated with a change of state without an accompanying change of temperature. Heat energy which causes a change of temperature without a change of state is referred to as *sensible heat*.

2.8 Crystals

Many substances, including metals, have a *crystalline* structure in the solid state, that is, their basic particles are arranged in definite three-dimensional patterns of rigid geometrical form. This pattern is repeated many times, provided the crystal is allowed to grow freely. Figure 2.5 shows some copper sulphate crystals and their regular geometric shape is

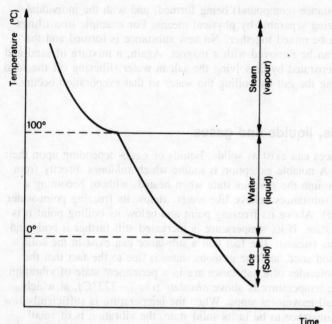

Fig. 2.4 Cooling curve for water

clearly apparent. Non-crystalline solids such as pitch, glass and many polymeric (plastic) materials do not have their basic particles arranged in a geometric pattern. Their particles have a random formation and as a result such substances are said to be *amorphous* (without shape).

The structure of crystals is better understood if their constituent atoms are considered to be spherical in shape. Figure 2.6(*a*) shows a simple cubic crystal built up from eight spherical particles. The broken lines joining the centres of the spheres represent the *unit cell* of this simple crystal. The unit cell is the geometric figure which illustrates the fundamental grouping of the particles in the solid. To form the crystal this unit cell is repeated many times to form the *space lattice* as shown in Fig. 2.6(*b*).

All crystal structures can be analysed into fourteen basic space lattices called the *Bravais space lattices*. For simplicity only the unit cells of each lattice are shown in Fig. 2.7. Of these fourteen possible space lattice formations, only six are met with in metal crystals. Of these a few of the most common are:

Body-centred cubic	*Face-centred cubic*	*Close-packed hexagonal*
Chromium	Aluminium	Beryllium
Molybdenum	Copper	Cadmium
Niobium	Lead	Magnesium
Tungsten	Nickel	Zinc

+----------Iron----------+

Fig. 2.5 Copper sulphate crystals growing from solution (J. Lewis)

2.9 Allotropy

Allotropy is the ability of a substance to exist in more than one physical form. The non-metal carbon is said to be allotropic since it can exist both as diamond (the hardest substance known to man) and as graphite (the soft 'lead' of a pencil). Both these substances consist solely of carbon atoms, but it is the crystal structure and the way in which the carbon atoms are bonded together where the difference lies. The metal iron is another allotropic material. Below 910°C iron has a body-centred cubic space lattice and is referred to as alpha (α) iron. Between 910°C and 1400°C iron has a face-centred cubic space lattice and is referred to as gamma (γ) iron. Above 1400°C iron again has a body centred cubic space lattice and is referred to as delta (δ) iron.

The allotropy of solids which relies solely on differences in their

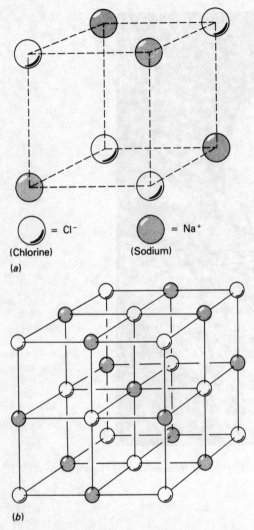

= Cl⁻
(Chlorine)

= Na⁺
(Sodium)

(a)

(b)

Fig. 2.6 The crystal structure (a) Unit cell for sodium chloride (common salt) crystal (b) Part of the space lattice for sodium chloride (eight unit cells shown)

crystal structure (space lattice) is referred to as *polymorphism*. Many metals are allotropic.

2.10 Grain structure

Although metals are crystalline solids, this is not immediately apparent when they are examined under the microscope. Figure 2.8(a) shows the

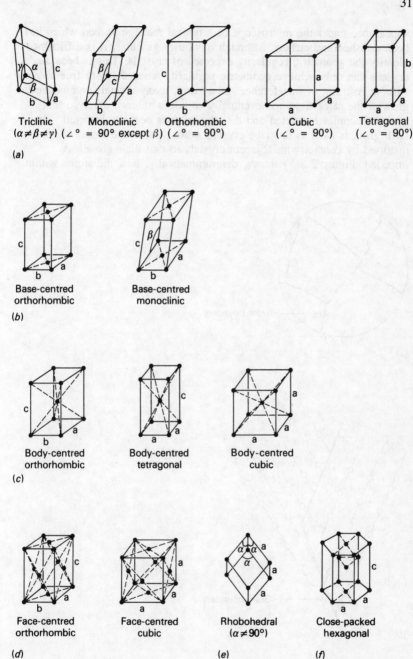

Fig. 2.7 Bravais space lattices (*a*) P-type (primitive space lattices) (*b*) C-type (base-centred on 'ab' face) (*c*) I-type (body-centred) (*d*) F-type (face-centred) (*e*) R-type (*f*) H-type

Triclinic
($\alpha \neq \beta \neq \gamma$)

Monoclinic
($\angle° = 90°$ except β)

Orthorhombic
($\angle° = 90°$)

Cubic
($\angle° = 90°$)

Tetragonal
($\angle° = 90°$)

(*a*)

Base-centred orthorhombic

Base-centred monoclinic

(*b*)

Body-centred orthorhombic

Body-centred tetragonal

Body-centred cubic

(*c*)

Face-centred orthorhombic

Face-centred cubic

Rhobohedral
($\alpha \neq 90°$)

Close-packed hexagonal

(*d*)

(*e*)

(*f*)

32

appearance under the microscope of a typical metal specimen which has been polished and etched. Although obviously granular, it is difficult to identify the geometric regularity expected of crystals. This is because crystals can only achieve geometric regularity when they are free to grow without restraint or interference. In a metal, many crystals commence to grow at the same time and eventually collide with each other so that growth becomes restricted and their boundaries become distorted. The term *grain* is used to describe crystals whose geometric shape has been distorted by contact with adjacent crystals so that their growth is impeded. Figure 2.8(*b*) shows, diagrammatically, how the atoms within a

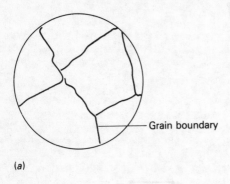

(*a*)

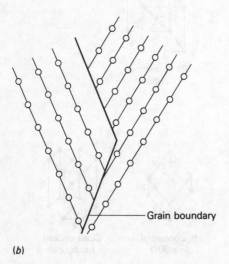

(*b*)

Fig. 2.8 Grain structure (*a*) Appearance of granular structure of metal under the microscope after etching (*b*) Despite the irregular appearance of the grain structure due to boundary interference, the crystal lattice within the grain is correctly ordered

grain can have the regular geometric space lattice expected of a crystal, and how that pattern breaks down at the grain boundary to make way for the geometric space lattice in an adjacent grain.

2.11 Crystal growth

When metals are in the liquid state, there is no orderly arrangement of the atoms which are free to move about with respect to each other (Section 2.7). Thus, in the liquid state, metals possess *mobility*. As the temperature of a molten metal falls, a point is reached where the metal starts to solidify. At this point the atoms change from a disordered or *amorphous* state to an ordered or *crystalline* state.

Like all pure crystalline substances, pure metals solidify at a fixed temperature as shown in Fig. 2.9(*a*). However, under industrial conditions, such purity will not be obtained and the crystal nucleus will form around an impurity particle such as a particle of slag. If the purity is of a high order, some under-cooling may occur before *nucleation* (the formation of the nucleus of the crystal) sets in as shown in Fig. 2.9(*b*). Amorphous (non-crystalline) solids such as glass, pitch and some polymers (plastics) exhibit no such change point and there is a gradual change from the solid state to the liquid state, as shown in Fig. 2.9(*c*).

Once the nucleus of the crystal forms, it provides a solid/liquid interface where crystallisation will proceed. For metals, the nuclei so formed will generally be face-centred cubic, body-centred cubic, or close-packed hexagonal unit cells. As the crystals grow they tend to develop spikes and change into 'tree-like' shapes called *dendrites* (Greek: *dendron* = a tree). Figure 2.10 shows a typical metal dendrite. The dendritic crystal grows until the spaces between the branches fill up. Growth of the dendrite ceases when the branches of one dendrite meet those of an adjacent dendrite. This process continues until eventually the entire liquid solidifies.

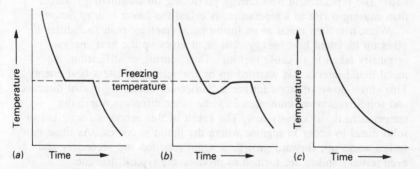

Fig. 2.9 Cooling curves (*a*) Pure metal: no undercooling (*b*) Pure metal: some undercooling (*c*) Amorphous solid: no single freezing temperature

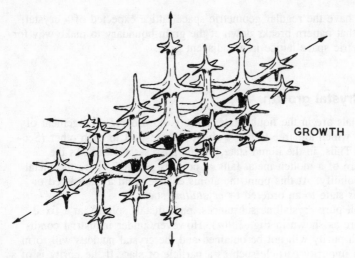

GROWTH

Fig. 2.10 Metallic dendrite growth (R.A. Higgins)

At this point there is little trace of the dendritic structure left, and it is only possible to see the grains into which the dendrites have grown. The steps in the growth pattern of a crystal from nucleus to grain are shown in Fig. 2.11.

The reason for dendritic growth is as follows. When a solid metal is heated, its atoms vibrate about fixed points called lattice points, each atom being held in place by forces of attraction. As the temperature of the metal increases the energy of each atom increases. At a certain temperature (depending upon the metal) the atoms become so agitated that they can escape from their fixed positions. Thus the structure begins to lose rigidity, that is, it begins to melt. In moving from their fixed positions the atoms and molecules do work against their binding forces and energy is used up. This energy is replaced by the heat source of the furnace. The replacement heat energy producing fusion (melting), rather than causing a rise in temperature, is called the *latent heat of fusion*.

When a molten metal at its fusion point (melting point) solidifies it gives up its latent heat energy, that is, it gives up the heat energy originally taken in to cause melting. Thus, during solidification, the metal/liquid interface is warmed up by the release of latent heat energy. This slows down or stops further solidification occurring in that direction, and solidification recommences in some other direction where the temperature is sufficiently low. The result of this action is for spikes of solid metal to occur in regions where the liquid is coolest. As these new spikes warm up, forward growth is again retarded and secondary and even tertiary spikes are formed to produce the typical dendrite.

Although it is hard to relate a dendrite to the well-ordered crystal structures previously considered, it must be remembered that the unit

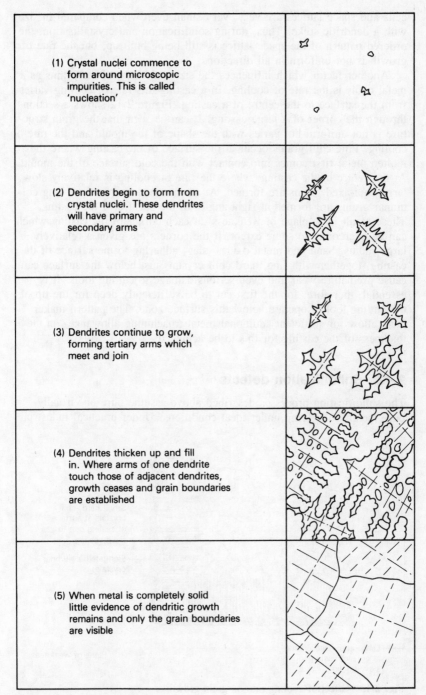

(1) Crystal nuclei commence to form around microscopic impurities. This is called 'nucleation'

(2) Dendrites begin to form from crystal nuclei. These dendrites will have primary and secondary arms

(3) Dendrites continue to grow, forming tertiary arms which meet and join

(4) Dendrites thicken up and fill in. Where arms of one dendrite touch those of adjacent dendrites, growth ceases and grain boundaries are established

(5) When metal is completely solid little evidence of dendritic growth remains and only the grain boundaries are visible

Fig. 2.11 Crystal growth

cells and space lattices are very, very small even when compared in size with a dendritic spike. Thus, during solidification and crystallisation, the ordered pattern of the space lattice is still being built up, but the rate of growth is not uniform in all directions.

Another factor which influences the size and shape of the grains as a metal cools is the rate of cooling. In a casting this rate of cooling varies from the surface to the centre of a casting. Figure 2.12 shows a section through the corner of a large casting. It can be seen that the grain structure is not uniform but varies with the shape of the mould and the rate of cooling. Fine chill grains occur at the surface of the casting where the molten metal first comes into contact with the cold surface of the mould. At the centre of the casting, where the rate of cooling is relatively slow, large equi-axed grains are formed. At an intermediate position, long columnar grains are formed at right-angles to the surface of the casting. These result in a 'plane of weakness' at each corner of the casting which can be overcome, to some extent, if the corners are given a relatively large radius. Sand and metal oxides (slag) adhering to the surface of the casting, together with fine, hard chill crystals just below the surface can cause premature wear and even serious damage to cutting tools. It is essential, therefore, for the first cut to be sufficiently deep for the tip of the cutting tool to operate below this surface zone. The pattern maker must allow for sufficient additional metal (machining allowance) on the thickness of the casting for this to be achieved.

2.12 Solidification defects

The solidification processes described above assume pure or virtually pure metals solidifying under ideal conditions. Under practical industrial

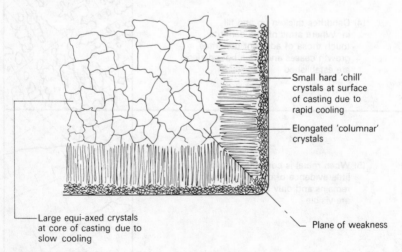

Small hard 'chill' crystals at surface of casting due to rapid cooling

Elongated 'columnar' crystals

Large equi-axed crystals at core of casting due to slow cooling

Plane of weakness

Fig. 2.12 Effect of cooling rate on grain growth

conditions impurities will inevitably be present. Also the dendrites will be growing in highly competitive conditions within the confines of the mould and each branch of each dendrite will be 'fighting' for space to develop. This overcrowding causes the dendritic branches or spikes to press against each other causing deformation. The more important solidification defects which can occur will now be considered.

Shrinkage

This can occur on two scales:

(a) very fine shrinkage leading to porosity between the dendrite branches;

(b) large scale shrinkage cavities resulting from poor mould design which prevents the feeding of molten metal to compensate for the normal volumetric shrinkage which occurs whenever metal solidifies and cools.

Shrinkage cavities occurring between the dendritic branches and spikes is known as *inter-dendritic porosity* and is caused by over-rapid cooling of the cast metal. As the molten metal between the dendrite branches and spikes cools and solidifies, it shrinks. Under normal cooling conditions additional molten metal has time to flow into the cavity so formed and no discontinuity occurs. If, however, cooling is too rapid there is not sufficient time for the shrinkage cavities to fill and porosity occurs along the dendrite branches. These fine cavities should not be confused with those due to gas porosity.

When a casting cools and solidifies there is always considerable volumetric shrinkage. In a well designed mould there are sufficient runners and risers (see Section 10.1) to feed additional molten metal back into the mould as shrinkage takes place. If the feeding of metal is inadequate during solidification *drawing* may occur. This is where the solidifying and shrinking metal in the smaller sections of the casting draws molten metal from adjacent, thicker sections of the casting instead of from the runners and risers. This leaves large cavities which weaken the casting and unsightly sunken surfaces which may not clean up during machining.

Misorientation

It has already been stated that dendrites form in a highly competitive environment and that their outermost branches and spikes are distorted as they interfere with each other at the grain boundaries. This results in the strict crystalline geometry of the space lattice breaking down at the grain boundary and producing an amorphous layer some two or three atoms thick as shown in Fig. 2.13. This amorphous layer behaves like a highly viscous liquid and allows slight movement to occur between the grains. This accounts for the creep which occurs in metals stressed over long periods of time, and why creep is greater at elevated temperatures when the viscosity of the amorphous layer is reduced. This also accounts for

38

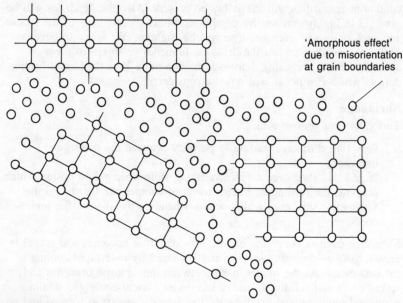

Fig. 2.13 Misorientation

the reason why metals tend to fracture by transverse cracking of the crystals at low temperatures when the amorphous layers are more viscous and resistant to inter-crystalline movement. At high temperatures failure is more likely to occur by inter-crystalline cracking than by transverse cracking since the amorphous layer is less viscous, weaker, and more likely to allow inter-crystalline movement.

Segregation and inclusions

Once solidification is complete, there will be no evidence of dendritic growth providing the metal is absolutely pure and no shrinkage cavities or inclusions are present. However such conditions never occur commercially and it is necessary to examine the effects of such inclusions. Two types of inclusions may be present and these will now be considered.

Dissolved inclusions

These tend to remain in the molten metal as long as possible so that the inclusions finally solidify in the spaces between the branches and spikes of dendrites along with the residual host metal. The presence of these dissolved inclusions often causes discolouration of the host metal outlining the shape of the original dendritic formation when examined under the microscope. Thus the dissolved inclusions are segregated from the host metal and are referred to as *minor segregations* as shown in Fig. 2.14. This figure also shows interdendritic porosity. Minor segregations are the most harmful since they lead to brittleness and cracking in the casting and

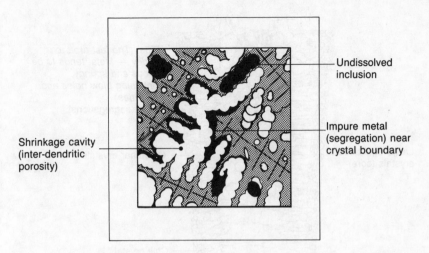

Fig. 2.14 Minor segregations and inter-dendritic porosity

may result, in the case of an ingot, in cracking and crumbling during subsequent forging and rolling. This is referred to as 'shortness' in the metal. Metal which is *hot-short* crumbles when hot worked, whilst metal referred to as *cold-short* crumbles when cold worked.

   Major segregations are only likely to be found in large castings such as ingots where large columnar crystals are present. As these move inwards they push the residual molten metal and dissolved inclusions ahead of them. Thus all the inclusions tend to become concentrated in the central 'pipe' where final solidification takes place. The appearance of these segregated inclusions in a sectioned and etched ingot casting is shown in Fig. 2.15.

Undissolved inclusions

These are substances such as sand particles washed from the sides of the mould by the molten metal and also metal oxide particles (scale). Generally such inclusions are less dense than the molten metal so they tend to float to the surface of the runners and risers where they can be discarded. However they sometimes become trapped in the casting where they form discontinuities from which fatigue cracks can originate. Such inclusions also adversely affect the strength and machining properties of the casting. A cutting tool hitting a hard inclusion will have its cutting edge blunted or chipped.

Gas porosity

Gases are frequently dissolved in the hot, molten metal but are expelled as the metal cools and solidifies. The sources of these gases may be from

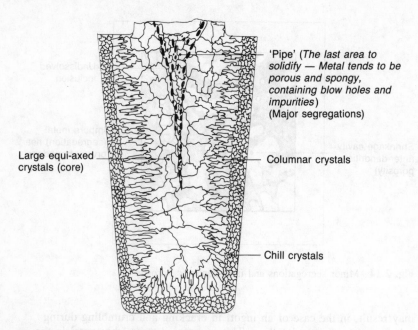

Large equi-axed crystals (core)

'Pipe' (*The last area to solidify — Metal tends to be porous and spongy, containing blow holes and impurities*) (Major segregations)

Columnar crystals

Chill crystals

Fig. 2.15 Major segregations

the furnace atmosphere, or from chemical reactions which take place during the melt. Most aluminium alloys and some copper alloys are susceptible to *gassing*. These metals tend to absorb hydrogen gas from the furnace atmosphere or from the moisture in the foundry atmosphere. The hydrogen is driven off by the addition of suitable chemicals immediately prior to pouring. This is known as *degassing*. Adequate ventilation is required during this operation as the gas driven off is in the form of hydrogen chloride which turns into hydrochloric acid on contact with atmospheric moisture. When ferrous metals such as steel and cast iron are cast, carbon monoxide gas may be present as the carbon in these metals tends to combine with the oxygen in the air. Any gases generated bubble out as the metal cools and become trapped between the branches of the dendrites to form small random cavities.

The moisture in green-sand moulds may boil to steam which will be trapped at or just below the surface of the casting to form larger blowholes if adequate venting of the mould is not provided. When exposed during machining, blow-holes appear as spherical or oval cavities with shiny surfaces.

2.13 Macro and microscopical examination

The grain structure of metals and any inclusions and discontinuities which may be present can be studied by macro or by microscopical examination.

Macro-examination implies the use of the unaided eye or the use of a low power magnifying glass. The sample component is sectioned and the surface is ground smooth. Since grinding tends to 'drag' the surface slightly, it is usual to hand finish the surface using a grade 0 or 00 abrasive paper. To reveal the grain structure it is necessary to *etch* the specimen. This is done by dripping a suitable etchant onto the surface of the specimen. The etchant will eat away the grain boundaries so that the individual grains stand out in relief. The specimen is then washed and examined. Details can frequently be seen more clearly whilst the surface is still wet. Table 2.1 lists a number of suitable etchants for macro-examination, whilst Fig. 2.16 shows a typical example of the appearance of a component prepared for macro-examination. A slag inclusion in the weld is just visible.

Microscopical examination requires much more careful preparation of the specimen. Since this has to be mounted on a microscope slide, it is usual to cut a small specimen from the component to be examined. The specimen is then mounted as shown in Fig. 2.17. Initial grinding must proceed with the utmost care to avoid overheating the specimen and altering its micro-structure. Intermediate finishing is then carried on progressively finer grades of abrasive paper. The paper is placed flat on a piece of plate glass and the specimen is moved back and forth so that the abrasive marks are a series of straight lines. At each change to a finer grade of abrasive the specimen is worked so that the new abrasive marks

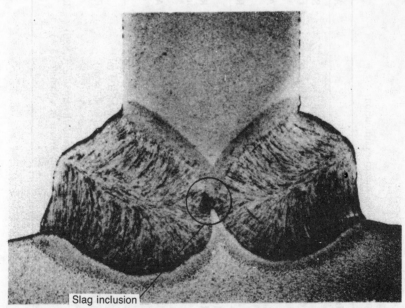

Slag inclusion

Fig. 2.16 Specimen as it appears for macro-examination

Table 2.1 Etchants for macro-examination

Material	Composition	Application
Steel	50% hydrochloric acid (conc) 50% water	Specimen boiled in etchant for 5 to 15 minutes. For revealing flow lines, structure of fusion welds, cracks, porosity, case depth.
	25% nitric acid (conc) 75% water	As above, but can be applied by cold swabbing for large specimens.
	Stead's Reagent	Reveals dendritic structure in steel castings, and phosphorous segregation.
Aluminium & Aluminium alloys	20% hydrofluoric acid (conc) 80% water	Reveals flow lines and general grain structure and impurities (undissolved inclusions).
	45% hydrochloric acid (conc) 15% hydrofluoric acid (conc) 15% nitric acid (conc) 25% water	As above, but more reactive reagent. Avoid contact with skin.
Copper & Copper alloys	25g ferric chloride in 100 ml 25% hydrocloric acid (conc) 75% water	Reveals dendritic structure of α phase solid solutions.
	33% ammonium hydroxide (0.880) 33% ammonium persulphate (5%) 34% water	Reveals β phase structure.

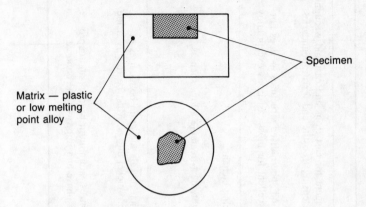

Fig. 2.17 Mounted specimen

are at right-angles to the previous ones. Treatment continues until the previous abrasive marks are no longer visible. Finally, the specimen is polished on a rotary metallurgical polishing machine. A suspension of jeweller's rouge or a suspension of diamantine is dripped onto the rotating pad of 'selvyt' cloth. Absolute cleanliness is essential to avoid scratching the polished surface and all traces of polishing agent must be removed before etching. Polishing continues until no marks can be seen on the unetched specimen under the microscope.

The etchant used and the method of application depends upon the metal being examined and the particular characteristics of the grain structure to be exposed. Microscopical metallurgy is a detailed study in its own right and only the very basic principles can be considered in this section. Table 2.2 lists a few of the more widely used etchants and their applications. Since metallurgical specimens are opaque, the metallurgical microscope uses reflected light. The light source is adjustable for intensity and colour filters are provided so that the light can be tinted to give the optimum contrast and clarity. Usually a turret of objective lenses is provided so that the magnification can be quickly and easily changed. Figure 2.18 shows the appearance of a typical etched specimen as seen under a microscope. The slag inclusion first identified in Fig. 2.16 is more easily visible under the microscope, as is the grain structure of the metal.

2.14 Polymeric materials

There is an ever increasing number of synthetic, polymeric materials available under the popular name of *plastics*. This is a misnomer since polymeric materials rarely show plastic properties in their finished condition. In fact many show elastic properties. The name 'plastics' comes from the fact that during the moulding process by which they are shaped,

Table 2.2 Etchants for microscopical examination

Material	Composition	Application
Steel	'Nital' 2% to 5% nitric acid (conc) 95% to 98% ethyl or methyl alcohol.	General purpose etching re-agent for cast irons and cast and wrought steels. Reveals pearlite, martensite, troostite and ferrite boundaries.
	'Picral' 4% picric acid (conc) 96% ethyl or methyl alcohol	Reveals details of pearlitic and spheroidised structures — does not attack ferrite boundaries. Excellent for most cast irons except alloy and ferritic cast irons.
Aluminium & aluminium alloys	0.5% hydrofluoric acid (conc) 99.5% water	General purpose cold-swabbing etchant.
	'Keller's Reagent' 1.0% hydrofluoric acid (conc) 1.5% hydrochloric acid (conc) 2.5% nitric acid (conc) 95% water	Immersion etching of wrought, heat-treatable alloys such as 'duralumin'.
Copper & Copper alloys	10g ammonium persulphate 20% ammonium hydroxide 80% water OR	General purpose etchant which reveals the grain boundaries of brasses, tin bronzes and cupro-nickel alloys. Must be freshly made before use.
	10g ferric chloride 35% hydrochloric acid (conc) 65% water	Suitable for $\alpha\beta$ brasses, bronzes, aluminium bronzes and cupro-nickel alloys. Darkens the β phase and improves the contrast.

Fig. 2.18 Specimen under microscopical examination

they are reduced to a plastic condition by heating them to just above the temperature of boiling water. There are three main groups of polymeric or 'plastic' materials.

Thermosetting plastics (thermosets)

This group of polymeric materials undergoes chemical change during the moulding process and can never again be softened by reheating. These materials are generally hard, rigid and rather brittle. A typical example is melamine formaldehyde used for making such articles as snooker and billiard balls and table-wear. The strength of thermosetting plastics can be greatly increased by reinforcing them with fibrous materials.

Thermoplastics

These become soft and can be remoulded each time they are reheated. They are not so rigid as thermosetting plastics and tend to be tougher. For example rigid polyvinyl chloride (PVC) is used for rain water gutter-ing and down-piping on buildings, and non-rigid PVC is used for the insulation of flexible electrical cables.

Elastomers

The elastomers, or rubbers, are cross-linked polymeric materials in which there are not sufficient cross links to make them as rigid as the thermosetting plastics, but just sufficient to make them return to their original dimensions when the deforming load is removed. Whereas thermosets show little elongation under stress, elastomers are capable of elongations of up to 1000 per cent at tensile failure. Elastomers are, therefore, capable of extreme elastic deformation at low levels of stress. Unlike metals, the strain is not proportional to stress for elastomers and this will be considered in greater detail in Section 11.7. Elastomers are usually polymerised as thermoplastics and then cross-linked (vulcanised) with sulphur at approximately every five-hundredth carbon atom. Increased vulcanisation increases the cross-linking and this, in turn, increases the stiffness and reduces the elongation properties of the material. Fully vulcanised, natural rubber becomes a rigid, brittle thermoset called 'ebonite'.

2.15 Polymer building blocks

The polymeric materials introduced in Section 2.14 are all built from carbon atoms in association with other elements such as oxygen, hydrogen, nitrogen, chlorine and fluorine. Carbon atoms have four chemical bonds or, as chemists would say, a valency of four. Hydrogen atoms have a valency of one, so if hydrogen and carbon are combined in the simplest way to give a molecule of methane (natural) gas, the molecule would appear as:

$$\begin{array}{c} H \\ | \\ H - C - H \\ | \\ H \end{array}$$

Thus four hydrogen atoms combine with one carbon atom to make one molecule of methane gas. This molecule is given in the chemical formula CH_4 and because it consists solely of hydrogen and carbon, it is referred to as a *hydrocarbon*. The hydrocarbons are found in crude oil, coal and natural gas. They can be classified into four main groups as follows.

Paraffins

These are the simplest of the four groups of hydrocarbons. They have a general formula of C_nH_{2n+2}. For example, in the methane molecule just considered there is only one carbon atom, so $n = 1$, and the number of hydrogen atoms is $2(1) + 2 = 4$. This agrees with the formula already stated as CH_4. One way of recognising paraffins is the fact that their names always end in -ane (as in methane, propane, octane, etc.). This is because the paraffins are *saturated* hydrocarbons or *alkanes*, that is, they contain the maximum number of hydrogen atoms in each case, as shown in Fig. 2.19, and this makes them rather inactive chemically. The paraffins are the most common group of hydrocarbons appearing in crude oil.

Formula	Name	Use
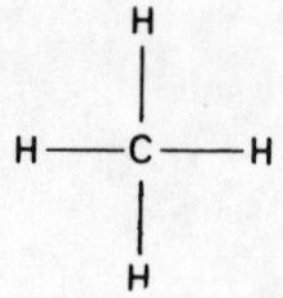 CH_4	Methane	Natural gas
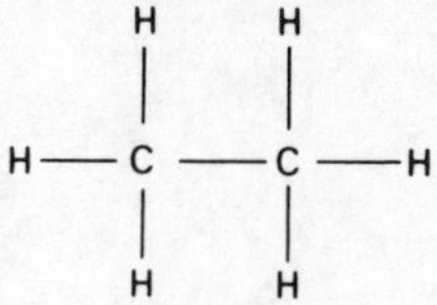 C_2H_6	Ethane	Converted into plastics
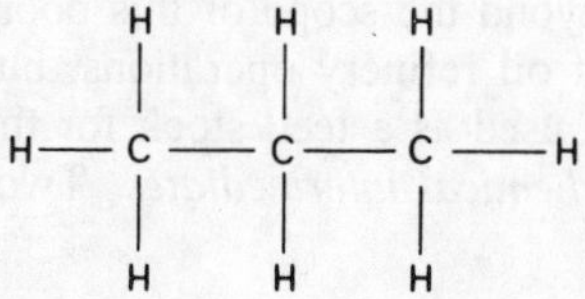 C_3H_8	Propane	Heating fuel
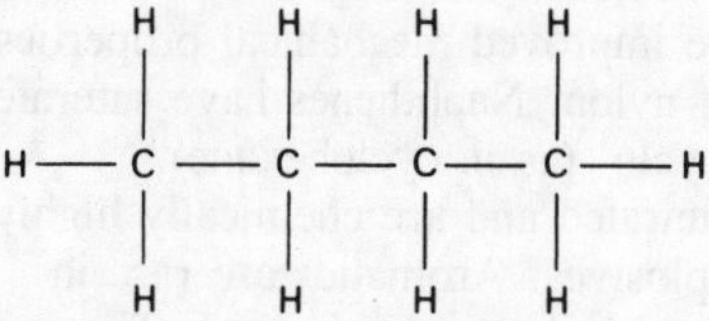 C_4H_{10}	Butane	1. Heating fuel 2. Converted into Synthetic rubbers

The series continues to C_{100} to become the asphalts and tars used for roads and roofing

H = Hydrogen.   C = Carbon.

Fig. 2.19 Common paraffins

Olefins

These are *unsaturated* hydrocarbons or *alkenes*, that is, additional hydrogen atoms have to be added to olefins in order to saturate them. This unsaturated condition makes them chemically reactive and olefins form the basis of many thermoplastic and elastomer materials. When their general formula is C_nH_{2n} they are called mono-olefins and are given names ending in -ylene (as in ethylene, propylene, etc.). There are more

Ethylene (derived from ethane)

Propylene (derived from propane)

H = Hydrogen
C = Carbon

Fig. 2.20 Common olefins

complex forms of the olefins but they are beyond the scope of this book. Olefins are usually produced in the course of oil refinery operations, but they are not abundant in crude oil. They are used as a feed stock for the polymer industry where they are known as *chemical intermediates*. Two typical examples are shown in Fig. 2.20.

Naphthenes and aromatics

These both have ring-shaped molecules, as shown in Fig. 2.21. Materials made from a ring-shaped molecule have improved mechanical properties, for example the high tensile strength of nylon. Naphthenes have saturated molecules and names beginning with 'cyclo' (as in cyclohexane). Aromatics, on the other hand, are unsaturated and are chemically highly reactive, being used in solvents and explosives. Aromatics are rare in crude oils (except those found in California), but occur in coal. They form the basis of the styrene group of plastics.

2.16 Polymers

Consider the manufacture of simple polymeric material such as polyethylene. The paraffin ethane is first converted into its corresponding olefin (hence the term 'chemical intermediate'). A single molecule of the olefin ethylene is referred to as a *monomer*, and the next stage of the process is to combine several monomers together to form a much larger molecule called a *polymer* (poly means many). In this example it is polyethylene. In the form of a polymer the olefin takes the characteristics of a plastic material. Two examples are shown in Fig. 2.22. There are some simple basic rules which govern the number of monomers which

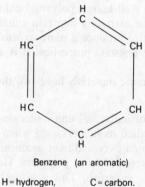

Cyclohexane (a naphthene)

Benzene (an aromatic)

H = hydrogen,   C = carbon.

Fig. 2.21 Common naphthenes and aromatics

can be brought together to form a polymer. For example, at room temperature ethylene which is made up of single molecules is a gas. A polymer of six monomers of ethylene is a liquid; a polymer of 36 monomers is a grease; a polymer of 140 monomers is a wax, and a polymer of 500 or more monomers is a solid plastic material. The upper limit is about 2000 monomers. At this point there is little further increase in strength, but a considerable increase in hardness and brittleness. This rule applies to most plastic materials.

Polymeric materials containing only carbon and hydrogen are highly flammable. To render these materials less flammable (and the rules governing building applications insist on this) at least one of the hydrogen

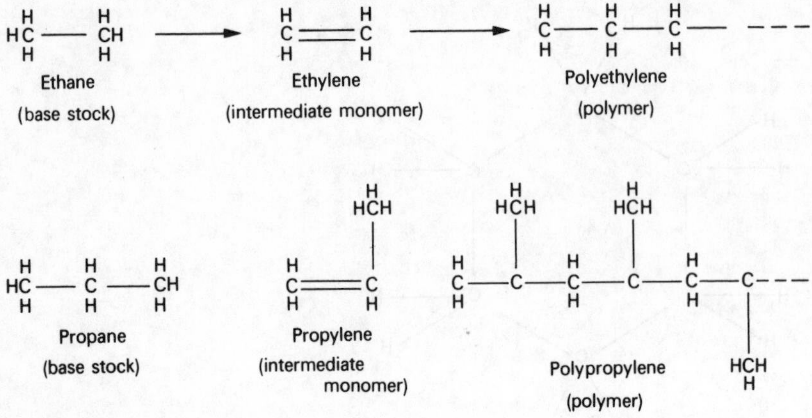

Fig. 2.22 Simple polymers

atoms in each monomer has to be replaced by a chlorine atom as shown in Fig. 2.23. The resulting polymer is the well-known polyvinyl chloride (PVC), a non-flammable plastic suitable for extruding into rain guttering. Fluorine may be added instead of chlorine to produce a more expensive material with superior mechanical and fire-resistant properties. It is also more resistant to sunlight.

Thus it becomes obvious that all polymeric materials have two things in common.

(a) They are all made up of long chains of individual unit molecules. These individual unit molecules are called monomers, and when large numbers of these monomers are repeated over and over again to form a long chain molecule they are referred to as polymers. Hence such materials are known as *polymeric materials*. This is a much more accurate description of these materials than the popular word plastic.

(b) They are all based on a chain of monomers which builds up a giant molecule. It is the shape of this chain as well as its composition which determines the properties of polymeric materials.

Figure 2.24 shows some of the shapes which the molecular chain of a polymeric material may take. The linear chain shown in Fig. 2.24(a) and the linear chain with side branches shown in Fig. 2.24(b) are typical of thermoplastic materials. The simple linear chains with no side branches can easily move past each other. This results in a non-rigid thermoplastic material which can be flexed and stretched. Such materials melt at low temperatures and easily return to their original state when they cool down. Polyethylene is an example of such a material. Since it is more difficult for branched linear chains to move past each other, materials with monomers in this configuration are more rigid, harder and stronger.

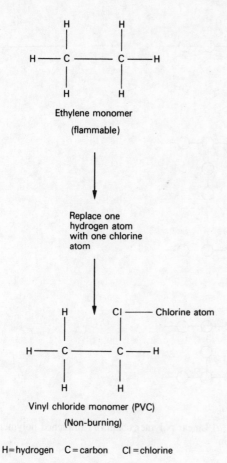

Ethylene monomer

(flammable)

Replace one
hydrogen atom
with one chlorine
atom

Cl ——— Chlorine atom

Vinyl chloride monomer (PVC)

(Non-burning)

H = hydrogen   C = carbon   Cl = chlorine

Fig. 2.23 Chlorinated plastics

Also they are less dense since the side branches prevent the chains being packed so closely together. Heat energy is required to break down the side branches and this raises the melting temperature above that for materials with a simple linear chain. An example of a thermoplastic material with a branched linear chain is polypropylene. The cross-linked molecular chain shown in Fig. 2.24(c) is typical of the thermosetting plastics. These are rigid and tend to be brittle once the links have been formed by *curing* the material during the moulding process. Once curing has occurred the material cannot be softened by reheating as the process is not reversible. If heated sufficiently they char or burn and are destroyed.

Thermosetting plastics differ from thermoplastic materials in the way in which polymerisation (curing) occurs. In thermoplastics, polymerisation

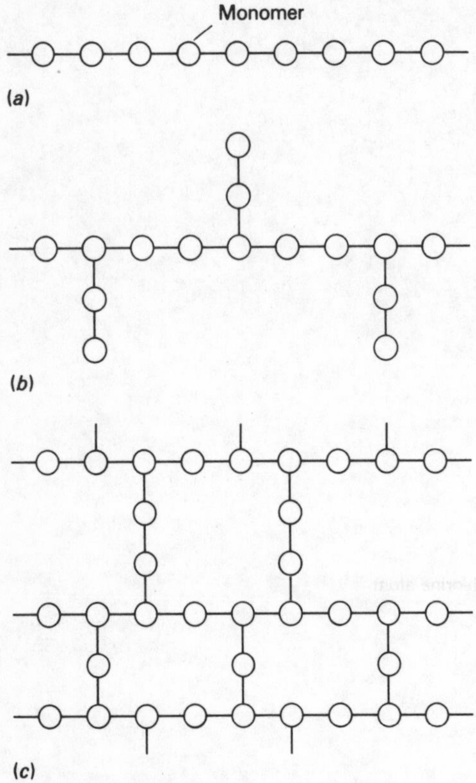

Fig. 2.24 Typical polymer chains (*a*) Linear polymer chain (*b*) Branched polymer chain (*c*) Cross-linked polymer chain

occurs through the addition of monomers at the time of manufacture and no further curing occurs during the moulding process. In thermosetting plastics, polymerisation usually occurs through condensation. In this latter process the plastic moulding material reacts within itself, or with some other chemical (the hardener), when heated to a critical temperature. At this temperature the moulding material releases or 'condenses' out small molecules such as water and polymerisation becomes complete. This loss of water results in volumetric shrinkage which has to be allowed for in the moulding process and, also, the moulds have to be designed with vents to allow the steam generated during polymerisation to escape. The principle of polymerisation by condensation is shown in Fig. 2.25.

2.17 Crystallinity in polymers

Crystals have already been described as having their particles arranged in recurring well-ordered geometric patterns. Materials which do not have

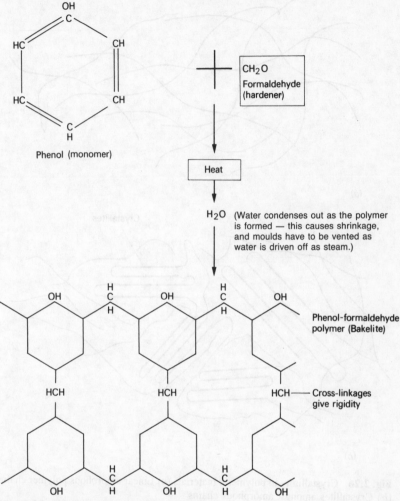

Fig. 2.25 Curing of thermosetting plastics

this ordered arrangement of geometric patterns as their basic structure have been described as amorphous. For example the polymeric material PTFE (used for coating non-stick cooking utensils) has a carbon chain which is helical with 14 atoms per turn of the helix and to which are attached side chains. It is hardly surprising that a polymer with the shape of a coil spring with side chains of attached methyl groups (CH_3), or with aromatic rings, has little chance of taking up the ordered patterns of a crystalline material. Hence most polymeric materials are amorphous or even glass-like (supercooled high viscosity liquids).

Such amorphous arrangements of polymer chains are often represented

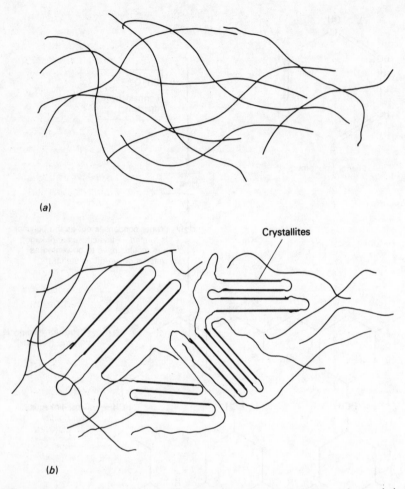

Fig. 2.26 Crystallinity in polymeric materials (*a*) Linear amorphous polymer chains (*b*) Crystallites amongst amorphous chains

by a tangle of lines as shown in Fig. 2.26(*a*). Each line represents an individual molecular chain, but does not show the individual atoms for the practical reason that there would be too many and they would be too small to draw. However, simple linear chains without side branches or cross links may show some degree of ordering on a sub-microscopic scale. Such ordered regions are called *crystallites* and there may be several such regions along a single molecular chain. Such an arrangement is shown in Fig. 2.26(*b*) where it can be seen that the individual molecular chains extend through several crystalline and non-crystalline regions.

The *crystallinity* of a polymeric material is defined as the ratio between the mass of the crystallites and the total mass of the material being considered. For example a material having 80 per cent crystallinity will consist of 80 per cent crystallite structure and 20 per cent non-crystallite (amorphous) structure. Since the monomers making up the polymer chain are packed more closely together in crystallites, it follows that materials with a high crystallinity will be more dense than materials with a low crystallinity. For example, low density polyethylene with a crystallinity of only 50 to 70 per cent will have a density of about 920 kg m^{-3} and a melting point of 115°C; whereas high-density polyethylene with 75 to 95 per cent crystallinity will have a density of about 950 kg m^{-3} and a melting point of 135°C. The crystallinity of a polymeric material has a marked effect upon its properties. For example, increasing the crystallinity of a material:

(a) increases its melting point and, instead of softening gradually with increased temperature, it will exhibit a sharper melting point which is similar to that of fully crystalline materials;

(b) increases the resistance of the material to the absorption of water and to solvent attack since it is more difficult for the water and solvent to penetrate the high-density crystallites than it is to penetrate the more open amorphous structure;

(c) prevents the penetration of plasticisers and this reduces the ultimate elongation of the material;

(d) makes the material more impervious to gases and this may be useful in food packaging and protective coatings. However, this high level of impermeability is a disadvantage in polymer fibres which must be coloured by dyeing.

The effect of crystallinity on the ultimate tensile strength and percentage elongation of a typical polymeric material is shown in Fig. 2.27. The relative crystallinity can be modified by heat treatment. For example a crystallisable polythene can be given a crystallinity of 80 per cent by slow cooling or a crystallinity of only 65 per cent by rapid cooling (quenching).

2.18 Orientation

The intermediate condition of *orientation* lies between the amorphous and the crystalline state. The processing of the plastic material and the effects produced are similar to those involved in the cold-working of metals. The mechanical working of amorphous and crystalline polymers has the following effects upon their morphology and properties.

Uniaxial orientation

If a polymer is drawn into fibres through a die (similar to metal wire drawing), the molecules and the crystallites of the polymer are aligned

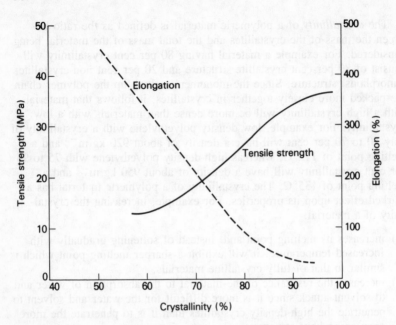

Fig. 2.27 Effect of crystallinity on the properties of polyethylene

parallel to the direction of drawing. That is, the structure has become oriented in the direction of drawing. This oriented condition greatly increases the tensile strength and the impact strength levels compared with the same polymer in the bulk condition.

To understand how this improvement in properties comes about, consider the typical stress-strain graph for a crystalline polymer as shown in Fig. 2.28. From O to A the material suffers elastic deformation and on removal of the stress the displaced atoms return to their original position. From the point A, plastic deformation occurs. Providing the strain rate (the speed at which the material is deformed) is low enough, two things will happen. Firstly the polymer chains will unfold and straighten out, and secondly they will slip over each other. Thus the polymer chains will end up aligned in the direction of drawing. They will also be packed together in a very orderly manner. This results in the increased tensile strength and toughness mentioned earlier.

Unfortunately, materials which have been subjected to uniaxial working (working in one direction only) only benefit from improved properties in the direction of working. They still remain relatively weak and tend to split when forces are applied at an angle to the direction of orientation. For fibres this is relatively unimportant since they are nearly always loaded in the direction of polymer chain orientation, however for films (sheet) this can be a serious defect.

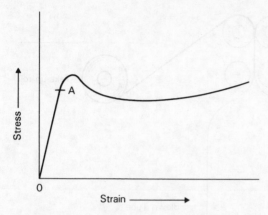

Fig. 2.28 Typical stress/strain curve for a crystalline polymer

Biaxial orientation

Films and sheets are produced so that the polymer chains are biaxially oriented, that is, the material is stretched longitudinally and transversely during production. This ensures uniform strength in whichever direction the film or sheet is stressed.

Film is the term used for flat material which is less than 0.25 mm in thickness. The process of film blowing is shown in Fig. 2.29. The plastic material is extruded vertically as a tube from an annular die. The wall thickness of the tube so produced is usually in the order of 0.4 mm to 0.6 mm. The tube is closed by pinch rolls high above the point of extrusion, and air is blown into the tube through the centre of the die mandrel to inflate the tube into a thin walled bubble. The wall thickness of the bubble is the final film thickness and stretching is uniform in all directions. Usually the bubble diameter is 1.5 to 3 times the die diameter. The tube is finally slit and opened out to make a flat film.

Sheet is produced by extruding the plastic material through a *sheeting die* as shown in Fig. 2.30(*a*) and then *calendering* the extruded sheet between rolls as shown in Fig. 2.30(*b*) to orientate the polymer chains, reducing the sheet to its final thickness (over 0.25 mm for sheet, under 0.25 mm for film) and imparting the required surface finish.

The optical properties of polyolefins are greatly improved by biaxial orientation. The best gloss and clarity is given by rolling, although the clarity of blown film improves as the blow ratio (ratio of bubble diameter to die annulus diameter) is increased.

2.19 Melting points of crystalline polymers (Tm)

Crystalline materials such as metals show well defined melting points when heated to sufficiently high temperatures. However, amorphous

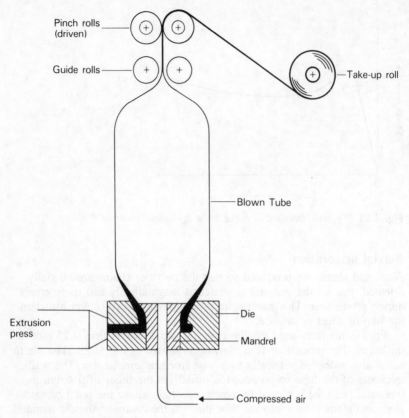

Fig. 2.29 Film blowing

solids do not show a clearly defined melting point when heated but merely become progressively less rigid until they eventually become liquid. Amorphous thermoplastic materials behave in this way. At room temperature they are so viscous that they behave as solids, but as the temperature rises the material becomes progressively less viscous until it becomes liquid without showing any clearly defined melting point.

Other thermoplastic materials show some crystallinity, and for these it is possible to determine a melting temperature (Tm). The melting temperature is determined by plotting the specific volume of the material against temperature rise as shown in Fig. 2.31. Initially the smaller, less perfect crystallites become amorphous and this is indicated by the portion of the curve marked (AB). At the point (B), a rapid increase in specific volume with temperature occurs. This point is regarded as the melting temperature (Tm), and is defined as the point where the material loses its crystallinity completely and becomes amorphous. The greater the crystallinity of any thermoplastic material, the higher will be its melting

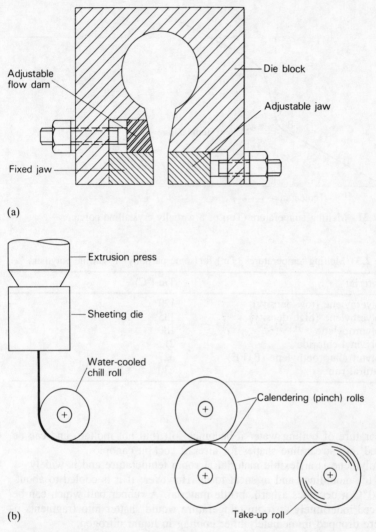

(a)

(b)

Fig. 2.30 Plastic sheet production (*a*) Sheeting die (*b*) Calendering

temperature. Table 2.3 lists the melting points for some typical crystalline polymers.

2.20 Glass transition temperature (Tg)

Polymethyl methacrylate (Perspex) is a rigid, glass-like plastic material with excellent optical properties at room temperature. At just above the

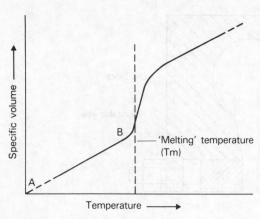

Fig. 2.31 Melting temperature (Tm) of a partially crystalline polymer

Table 2.3 Melting temperatures (Tm) for some partially crystalline polymers

Material	Tm (°C)
Polyethylene (low density)	120
Polyethylene (high density)	135
Polypropylene	180
Polyvinyl chloride	212
Polytetrafluoroethylene (PTFE)	327
Natural rubber	30

temperature of boiling water it becomes soft (but not molten) and can be formed into streamline shapes for aircraft cockpit canopies.

Polythene is a flexible material at room temperature and is widely used for mouldings and in sheet form. However, if it is cooled to about −120°C it becomes a hard, brittle material. A rubber ball which can be bounced indefinitely at room temperature would shatter into fragments if it were dropped immediately after cooling in liquid nitrogen.

The temperature at which a polymeric material changes from being rigid and brittle to being flexible and rubbery is called the *glass transition temperature* (Tg). The glass transition temperature (Tg) is less well defined than the melting temperature (Tm) and is difficult to determine. However, at the glass transition temperature, the tensile modulus (see Section 11.7) undergoes an abrupt change as shown in Fig. 2.32, and this can be used to determine the glass transition temperature. Below the glass transition temperature polymeric materials show a relatively high tensile modulus, with little extension and a high level of rigidity. Above the glass transition temperature, the tensile modulus is lower, the level of rigidity is lower, and the extension is very considerably increased as

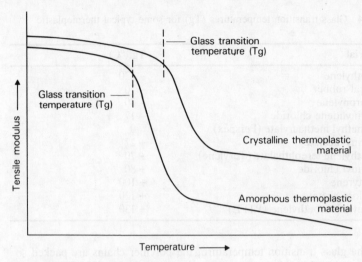

Fig. 2.32 Glass transition temperature (Tg)

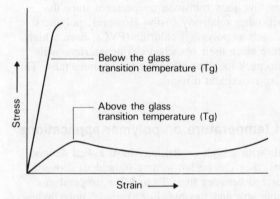

Fig. 2.33 Effect of the glass transition temperature (Tg) on the mechanical properties of a typical thermoplastic material

shown in Fig. 2.33. The glass transition temperature varies widely from one polymeric material to another, as can be seen from Table 2.4.

The reason for the change in properties at the glass transition temperature is as follows. Above the glass transition temperature the polymer chains for thermoplastic materials are reasonably free to move about so that, when stressed, they can uncoil and slide over each other. When this can happen to the polymer chain the material is soft, elastic and tough. However, as the temperature is lowered, the density of the material increases as the polymer chains pack more closely together.

Table 2.4 Glass transition temperatures (Tg) for some typical thermoplastic materials

Material	Tg (°C)
Polyethylene	– 120
Natural rubber	– 70
Polypropylene	– 30
Polyvinylidene chloride	– 17
Polymethyl methacrylate (Perspex)	0
Polyvinyl acetate	+ 27
Polyethylene terephthalate (Terylene)	+ 70
Polyvinyl chloride	+ 80
Polystyrene	+ 100
Cellulose acetate	+ 120
Polytetrafluoroethylene (PTFE)	+ 130

Below the glass transition temperature, the polymer chains are packed so closely together that relative movement becomes extremely difficult and little extension occurs, thus the material becomes hard and brittle. Polymers with linear molecules with no side branches, such as polyethylene, have a very low glass transition temperature since the chains can slide over each other relatively easily. However, polymers having branched chains, such as polyvinyl chloride (PVC), have a high glass transition temperature since their branched chains interfere with each other as they start to pack together when the temperature falls. This interference makes relative movement difficult.

2.21 The effect of temperature on polymer applications

Since polymeric materials with a high crystallinity have a well defined melting temperature (Tm), they can be hot-formed (moulded) above this temperature, and cold-formed between their Tg and Tm temperatures when they will be solid but soft and flexible. For example, polyethylene with a 95 per cent crystallinity has a Tg of − 120°C and a Tm of 138°C. Thus it is soft and flexible over a wide range of temperatures. The maximum service temperature is usually taken as approximately 85 per cent of the melting temperature, which in this case is 120°C. Compare this with a polythene with only 60 per cent crystallinity where the Tm is reduced to 115°C and a service temperature of only 98°C. The glass transition temperature is unaffected by the change in crystallinity.

Amorphous polymeric materials are usually moulded or formed above their glass transition temperature where they are soft and pliable, but used below this temperature where they are rigid. For example rigid PVC is an amorphous polymer with a Tg of 87°C. It is normally softened by hot air blast or radiant heat before manipulating to shape. Since amorphous plastics do not have a well defined melting temperature, the service temperature is taken as 85 per cent of the glass transition temperature which, for rigid PVC, is 70°C.

2.22 Memory effects

Polymeric materials with uniaxial and biaxial polymer chain orientation can have their crystallinity restored by heat treatment. This heat treatment consists of heating the material between its Tg and Tm temperatures so that the polymer chains lose their orientation and recover their original amorphous or their original semi-crystalline structure. This ability to return to the prestretched, disoriented state is referred to as the *memory effect* of the material.

Use is made of the memory effect in the packaging industry by heat shrinking protective foil around prepacked food and other commodities. The commodity to be packed is wrapped loosely in the foil which is then heated to above its glass transition temperature but well below its melting temperature. This causes restoration of the original disoriented polymer structure and shrinkage occurs resulting in the commodity becoming tightly packed.

3 Alloying of metals

3.1 Alloys

Pure metal objects are used where good electrical conductivity, good thermal conductivity, good corrosion resistance or a combination of these properties are required. However, pure metals usually lack the strength required for structural materials and alloys are used to give superior mechanical properties. For example, properties such as tensile strength, yield strength and hardness are improved by alloying but ductility is reduced. Since alloys can be designed to give specific properties they can be 'tailored', to suit a particular application.

An alloy is an intimate association of two or more component materials which form a single metallic liquid or solid. The component materials may be metal elements, or they may be metal elements and non-metal elements, or they may be metal elements and chemical compounds. It is important to distinguish between alloying elements and impurities. *Alloying elements* are deliberately added in controlled quantities to modify the properties of a material to match a particular specification. *Impurities* are undesirable and are usually carried over from some previous process such as smelting or casting. Since they impair the properties of the material, steps are usually taken to remove the impurities or reduce them to a level where their effects become insignificant.

Useful alloys can only be produced from component materials which are soluble in each other in the molten state, that is, they must be completely *miscible*. It would be useless to try to form an alloy from zinc and lead. The molten zinc would float on top of the molten lead and, upon cooling, they would form separate layers in the solid state with only tenuous bonding at the interface. Alloys are formed in three ways.

(a) If the alloying components in the molten solution have similar chemical properties, and their atoms are of similar size, they will not react together but will form a solid solution on cooling.

(b) If the alloying components in the molten solution have different chemical properties they may attract each other and form chemical compounds. Where the alloying components are both metals, these compounds are referred to as intermetallic compounds. Upon cooling the crystals will consist of a mixture of such compounds.

(c) In a situation where atoms with different chemical properties attract each other less than those with similar chemical properties, then both intermetallic compounds and solid solutions will be present at the same time. Upon cooling they will tend to separate out at the grain boundaries to form a heterogeneous mixture.

In any alloy the metal which is present in the larger proportion is referred to as the *parent metal* or *solvent*, whilst the metal (or non-metal) present in the smaller proportion is known as the *alloying component* or *solute*. Commercial alloys often contain more than one alloying element. For example, 'gun-metal' bronze alloy contains both tin and zinc in addition to the parent metal copper, whilst phosphor-bronze contains both tin and phosphorous in addition to the parent metal copper.

3.2 Solubility

In order to understand the formation of alloys, it is first necessary to understand the principles of solubility in the liquid and solid states. Sodium chloride (common table salt) dissolves readily in cold water. At room temperature, approximately 35 g of sodium chloride will dissolve in 100 g of water. The exact amount will depend upon the temperature of the water. If more sodium chloride is added to the solution it will not dissolve because the solution has already taken up all the salt it can dissolve and is said to be *saturated*. The excess salt will remain as a residue. The solubility of sodium chloride increases only slightly as the temperature of the water increases. In this example:

(a) the water is the solvent;
(b) the sodium chloride is the solute;
(c) the resulting liquid is the solution.

Figure 3.1 shows the difference between complete and partial solubility. The salt copper sulphate can also be dissolved in water, but unlike sodium chloride, its solubility increases substantially as the temperature of the water increases. This is shown in Fig. 3.2, where point A represents 50 g of copper sulphate being dissolved in 100 g of water.

66

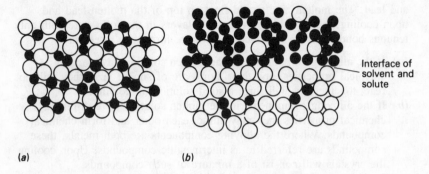

Fig. 3.1 Solubility (a) Complete solubility (b) Partial solubility

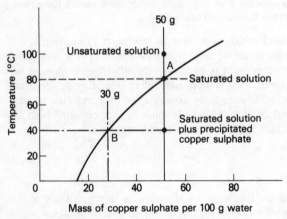

Fig. 3.2 Solubility curve for copper sulphate

(a) Above 80°C the water is capable of dissolving more than 50 g of copper sulphate, so the solution is said to be unsaturated.

(b) At 80°C the water will dissolve a maximum of 50 g of copper sulphate so the solution is said to be saturated.

(c) Below 80°C the water dissolves less than 50 g of copper sulphate. For example, at 40°C, (point B) only 30 g of copper sulphate can be dissolved in 100 g of water (can be 'held in solution') and the balance of 20 g of copper sulphate will be precipitated out of solution as a solid residue.

Substances which will not dissolve in a solvent are said to be insoluble. However, a substance which is insoluble in one solvent may be soluble in a different solvent.

3.3 Solid solutions

Most metals are completely and mutually soluble (they are miscible) in the liquid state, that is, when they are molten. Some, such as copper and nickel, not only form solutions in the molten or liquid state but remain in solution upon cooling and solidifying to become solid solutions. There are two sorts of solid solutions:

(a) *substitutional* solid solutions;
(b) *interstitial* solid solutions.

The copper-nickel alloy mentioned previously is an example of a substitutional solid solution. The more important factors governing the formation of a substitutional solid solution are as follows.

(a) *Atomic size.* The atoms of the solute and the solvent must be approximately the same size. If the atom diameters vary by more than 15 per cent the formation of a substitutional solid solution is highly unlikely.
(b) *Electrochemical series.* If there is only a small difference in charge between the alloying components then they will probably form a solid solution. Conversely if their charges are very dissimilar they are more likely to form intermetallic compounds.
(c) *Valency.* A metal of lower valency is more likely to dissolve one of higher valency than the other way round, assuming the conditions in (a) and (b) are also favourable. This holds good particularly for monovalent metals such as copper, silver and gold.

Figure 3.3 shows that both copper and nickel form face-centred cubic crystals. When these two metals are in solid solution they form a single face-centred cubic lattice with atoms of nickel replacing atoms of copper in the lattice. Hence the term substitutional solid solution. The substitution can be ordered, with the solute atoms taking up regular fixed positions of geometric symmetry in the lattice. However, most solid solutions are disordered, with the solute atoms appearing virtually at random throughout the solvent lattice.

Interstitial solid solutions are formed when the solute atoms are small enough to lie between the solvent atoms as shown in Fig. 3.4. For example, carbon atoms can form an interstitial solid solution with face-centred cubic crystals of iron.

3.4 Intermetallic compounds

It has already been stated that where the components of the alloy are sufficiently different chemically, they will tend to form compounds rather than solid solutions. In general, intermetallic compounds tend to be hard and brittle and are thus less useful for engineering alloys than the tough and ductile solid solutions. Intermetallic compounds are most widely

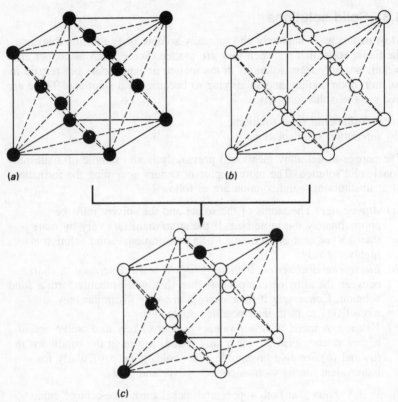

Fig. 3.3 Substitutional solid solution (*a*) Face-centred cubic crystal of copper (*b*) Face-centred cubic crystal of nickel (*c*) Substitutional solid solution of copper and nickel

found in bearing alloys where they form hard, wear-resistant pads with a low coefficient of friction, set in a matrix of tough, ductile solid solution.

3.5 Cooling curves

Most substances can exist as gases, liquids and solids, depending upon their temperature. Water is one such substance, which can exist as a gas or vapour (steam) if it is sufficiently hot; as a liquid, and as a solid (ice) if sufficiently cold. If water is raised to its boiling point and allowed to cool slowly, the change in temperature with time can be plotted as a graph as shown in Fig. 3.5. Such a graph is called a cooling curve. It can be seen from the graph that where a change of state occurs (such as liquid water to solid ice) there is a short pause in the cooling process. This pause is referred to as an *arrest point* and is the result of the water giving up latent heat energy as it changes into ice. Latent heat is the heat

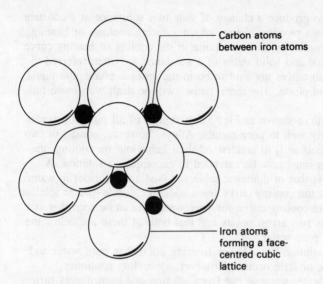

Fig. 3.4 Interstitial solid solution

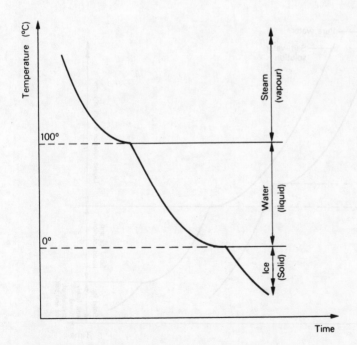

Fig. 3.5 Cooling curve for water

energy required to produce a change of state in a substance at a constant temperature. Thus a physical change of state during cooling, or heating, is always accompanied by an arrest point in the cooling or heating curve. The gaseous, liquid and solid states of a substance are often referred to as *phases*, and substances are said to be in the gaseous phase, the liquid phase, or the solid phase. The term 'phase' will be dealt with more fully in Section 3.6.

The cooling curve shown in Fig. 3.5 is typical of all pure substances and applies equally well to pure metals. Alloys, however, consist of two or more components and, to understand their behaviour on cooling, the above explanation must now be extended to encompass a solution. A suitable solution is that of domestic table salt (sodium chloride) in water. Figure 3.6 shows the cooling curve for a sodium chloride–water solution compared with the cooling curve for pure water. It can be seen that salt water solution has two arrest points and that both of these are below the freezing point of pure water.

A salt-water solution has a lower freezing point than pure water and at 0°C no change of state occurs. However, as cooling continues, droplets of pure water separate out from solution and immediately turn

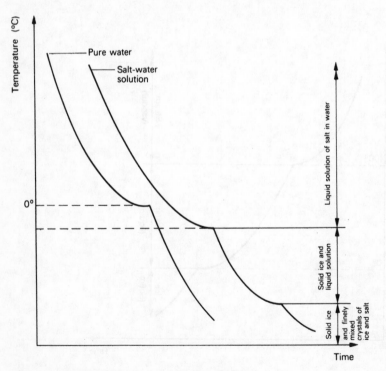

Fig. 3.6 Cooling curve for a salt-water solution

into ice particles. This occurs at the upper arrest point, which is usually not too well defined. The process of separation continues as the temperature of the remaining solution is further reduced. Thus as the temperature continues to fall more and more, water separates out and freezes, causing the concentration of the remaining salt water to increase. When the lower arrest point is reached, even the concentrated salt-water solution freezes and no liquid phase is left. The solid so formed consists of a mixture of fine crystals of pure water (ice) and fine crystals of salt (sodium chloride).

If the experiment is repeated several times using stronger and weaker salt-water solutions, a family of cooling curves can be plotted on the same axes as shown in Fig. 3.7. Reference to this figure shows that:

(a) the temperature of the lower arrest point remains constant;
(b) the temperature of the upper arrest point falls as the concentration of the solutions increases until a point is reached where the temperatures of the upper and lower arrest points coincide;
(c) the ratio of solid to liquid where the temperatures of the arrest points coincide is referred to as a *eutectic* composition. Solutions with a lower concentration of solid to liquid are referred to as *hypo-eutectic* solutions. Solutions with higher concentrations of solid to liquid than that of the eutectic composition are referred to as *hyper-eutectic* solutions.

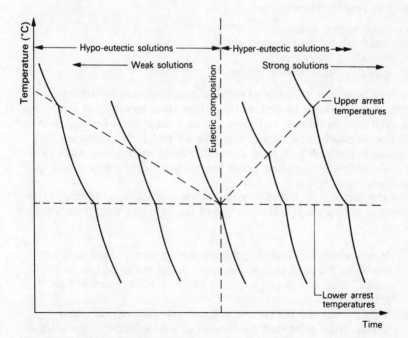

Fig. 3.7 'Family' of cooling curves

(*d*) When the concentration of the solution increases beyond that of the eutectic composition the temperature of the upper arrest point rises once more.

Since the amount of salt which can be held in solution with water varies with temperature, water separates out as ice crystals between the arrest points of hypo-eutectic solutions as the temperature falls, and salt crystals separate out between the arrest points of hyper-eutectic solutions as the temperature falls. Therefore the remaining solution is always of a constant concentration and has a constant arrest point temperature. This concentration is the eutectic composition. The fact that excess water or salt is rejected from solution so that a eutectic 'balance' is always ultimately achieved results in the diagram formed from the cooling curves (Fig. 3.7) being referred to as a *phase equilibrium diagram*. (Also referred to as a 'thermal equilibrium diagram' in some older texts.)

3.6 Phase

The term 'phase' has already been introduced in the previous section and it is now necessary to consider it in more detail. A phase may be defined as: *'a portion of a system which is of uniform composition and texture throughout, and which is separated from the other phases by clearly defined surfaces'*. Thus for the salt-water solution just considered there are four possible phases:

(1) water vapour (steam);
(2) liquid salt solution (sodium chloride in water);
(3) crystals of water (ice);
(4) crystals of salt (sodium chloride).

Each of these four phases is of uniform composition and is separated from adjacent phases by definite boundaries. It is necessary to distinguish between crystals (grains) and phases, since each phase is homogeneous but not necessarily continuous. Thus, the ice phase may appear as separate lumps with each lump containing many single-phase water (ice) crystals and the single-phase crystalline sodium chloride may appear as many separate crystals.

Extending this argument to metal alloys, when a liquid solution of two metals (a *binary* alloy) solidifies, one of the following conditions will occur.

(*a*) Metals which are soluble in the liquid state may become totally insoluble in the solid state and separate out as grains of two pure metals. Thus two phases each consisting of many grains will be present.
(*b*) Metals which are soluble in the liquid state may remain totally soluble in the solid state resulting in a 'solid solution'. Thus a single phase consisting of many grains will be present.

(c) The two metals may react together chemically to form an 'inter-metallic compound'. Again a single phase consisting of many grains will be present.

Therefore a binary alloy may be built up in a number of different ways and may consist of

(a) two pure metals existing entirely separately in the structure. In prac-tice this is extremely rare since there is usually some solubility of one metal in another;
(b) a single solid solution of one metal dissolved in another;
(c) a mixture of two solid solutions if the metals are only partially soluble in each other;
(d) an intermetallic compound and a solid solution mixed together.

The individual grains found in any of these phases may vary considerably in size. Some are large enough to see with the unaided eye, whilst others are so small that a high powered microscope is required. Note that although the phases found in alloys, as described above, are formed from two metals, they may equally well be formed between a metal and a non-metal. For example, austenite is a solid solution of carbon in iron, whilst cementite is the compound iron carbide.

3.7 Alloy types

As has already been stated, alloys consisting only of two component metals are referred to as binary alloys. Even when more than two com-ponents are present, a lot of useful information can be obtained from a study of the binary diagram of the two principal components present. The constituent components of most commercially available binary alloys are completely soluble in each other in the liquid (molten) state and, in general, do not form intermetallic compounds, (the exceptions being some bearing metals). However, upon cooling into the solid state, binary alloys can be classified into the following types.

Simple eutectic type
The two components are soluble in each other in the liquid state, but are completely insoluble in each other in the solid state.

Solid solution type
The two components are completely soluble in each other both in the liquid state and in the solid state.

Combination type
The two components are completely soluble in the liquid state, but are only partially soluble in each other in the solid state. Thus this type of alloy combines some of the characteristics of both the previous types, hence the name 'combination type' phase equilibrium diagram.

74

These three types of binary alloy systems and their phase equilibrium diagrams will now be considered in greater detail.

3.8 Phase equilibrium diagrams (eutectic type)

Figure 3.8 shows a eutectic-type of phase equilibrium diagram. It can be seen that it is identical with the diagram produced for a sodium chloride and water solution (Fig. 3.7), that is, total solubility of the salt in water in the liquid state and total insolubility (crystals of ice and separate crystals of salt) in the solid state. In the general case of Fig. 3.8, the two components present are referred to as metal A and metal B. Although they are mutually soluble in the liquid state, both components retain their individual identities of crystals of A and crystals of B in the solid state.

Reference to Fig. 3.8 shows that the line joining the points where solidification begins is referred to as the *liquidus* and the line joining the points where solidification is complete is referred to as the *solidus*.

This type of equilibrium diagram gets its name from the fact that at one particular composition (E), the temperature at which solidification commences is a minimum for the alloying elements present. With this composition the liquidus and the solidus coincide at the same temperature, thus the liquid changes into a solid with both A crystals and B crystals forming instantaneously at the same temperature. This point on the diagram is called the eutectic, the temperature at which it occurs is the eutectic temperature, and the composition is the eutectic composition.

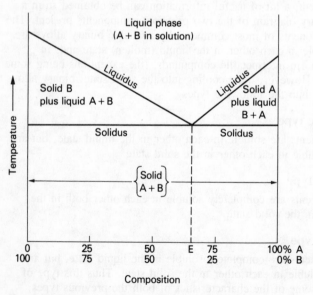

Fig. 3.8 Phase equilibrium diagram (eutectic type)

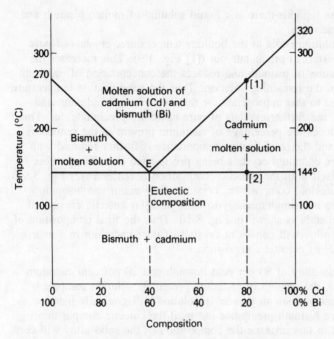

Fig. 3.9 Cadmium-bismuth phase equilibrium diagram

In practice, few metal alloys form simple eutectic type phase equilibrium diagrams. Exceptions to this are the cadmium-bismuth alloys and the phase equilibrium diagram for such alloys is shown in Fig. 3.9. It can be seen that the eutectic composition occurs when the alloy consists of 40 per cent cadmium and 60 per cent bismuth. For this composition solidification occurs at just over 140°C with both metals crystallising out of solution simultaneously. The eutectic structure is usually *lamellar* in form as shown in Fig. 3.10. In this instance there are alternate layers or 'laminations' of cadmium and bismuth.

Consider the cooling of an alloy consisting of 80 per cent cadmium and 20 per cent bismuth (a hyper-eutectic alloy).

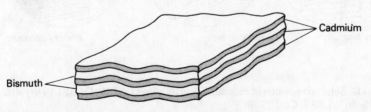

Fig. 3.10 Lamellar structure of eutectic composition

76

(a) Above the liquidus there is a liquid solution of molten bismuth and molten cadmium.

(b) As the solution cools to the liquidus temperature, crystals of pure cadmium start to precipitate out ([1] Fig. 3.9). This increases the concentration of bismuth and reduces the concentration of cadmium present in the remaining solution. Thus the solidification temperature is reduced to that appropriate for this new ratio of cadmium and bismuth, and further crystals of pure cadmium precipitate out. This again reduces the percentage of cadmium present in the remaining solution and the solidification temperature is further reduced with more pure cadmium crystals being precipitated out. This process repeats itself until the eutectic composition is reached ([2] Fig. 3.9).

(c) At the eutectoid composition, crystals of cadmium and bismuth precipitate out simultaneously to form lamellar eutectic crystals of the two metals as shown in Fig. 3.10. Thus the final composition of the solid alloy will consist of crystals of pure cadmium in a matrix of crystals of eutectic composition.

Similarly for an alloy of 80 per cent bismuth and 20 per cent cadmium (hypo-eutectic), the amount of cadmium present in solution compared with the amount of bismuth present in solution will gradually increase as crystals of pure bismuth precipitate out until the eutectic composition is reached. Thus in this instance the composition of the solid alloy will con-sist of crystals of pure bismuth in a matrix of crystals of eutectic composition.

For an alloy of 60 per cent bismuth and 40 per cent cadmium only crystals of eutectic composition will be present. These solid alloy com-positions are shown in Fig. 3.11.

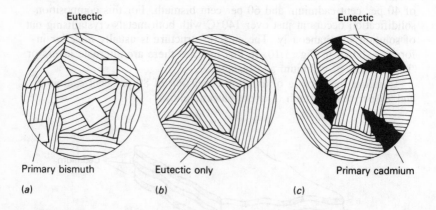

Fig. 3.11 Solid composition of cadmium-bismuth alloys (a) 20% Cd 80% Bi (b) 40% Cd 60% Bi (c) 80% Cd 20% Bi

3.9 Phase equilibrium diagram (solid solution type)

It has already been stated that copper and nickel are not only mutually soluble in the liquid (molten) state, they are also mutually soluble in the solid state and they form a substitutional solid solution. The phase equilibrium diagram for copper nickel alloys is shown in Fig. 3.12. Again, the line marked liquidus joins the points where solidification commences, whilst the line marked solidus joins the points where solidification is complete. This time there is no eutectic composition.

Thus for 100 per cent copper and 0 per cent nickel (pure copper) there is a single solidification temperature of 1084°C. This is to be expected since for a pure metal (in fact for any pure crystalline substance) the transition from liquid to solid takes place at a constant temperature. For an alloy of 80 per cent copper and 20 per cent nickel, Fig. 3.12 shows that solidification starts at 1190°C and is complete at 1135°C. Between the solidus and the liquidus is a solution of molten copper and nickel together with crystals of a solid solution of copper and nickel. For an alloy of 80 per cent nickel and 20 per cent copper Fig. 3.12 shows that solidification starts at 1410°C and is complete by 1380°C. Finally, Fig. 3.12 shows that 100 per cent nickel and 0 per cent copper (pure nickel) solidifies at the single temperature of 1445°C. Below the solidus the alloy consists entirely of crystals of copper and nickel in solid solution.

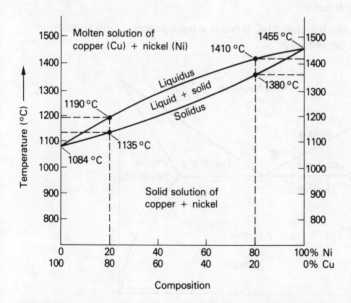

Fig. 3.12 Copper-nickel phase equilibrium diagram

78

3.10 Phase equilibrium diagram (combination type)

Many metals and non-metals are neither completely soluble in each other in the solid state, nor are they completely insoluble. Therefore they form a phase equilibrium diagram of the type shown in Fig. 3.13. In this system there are two solid solutions labelled α and β. The use of the Greek letters α, β, etc., in phase equilibrium diagrams may be defined, in general, as follows:

(a) a solid solution of one component A in an excess of another component B such that A is the solute and B is the solvent is referred to as solid solution α;

(b) a solid solution of the component B in an excess of the component A so that B now becomes the solute and A becomes the solvent is referred to as solid solution β;

(c) in a more complex alloy, any further solid solutions or inter-metallic compounds which may be formed would be referred to by the subsequent letters of the Greek alphabet. That is, γ, δ, etc.

Tin-lead alloys (soft solders) are a typical example of the combination type of phase equilibrium diagram as shown in Fig. 3.14. Reference to the tin-lead phase equilibrium diagram shows that the α phase is a solid solution of 19.2 per cent tin in 80.2 per cent lead at the eutectic temperature, and that the β phase is a solid solution of 2.6 per cent lead in 97.4 per cent tin at the eutectic temperature. This diagram can be explained as follows.

(a) Above the liquidus ABC there is a homogeneous solid solution of molten tin and lead.

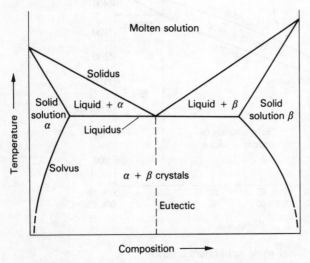

Fig. 3.13 Combination type phase equilibrium diagram

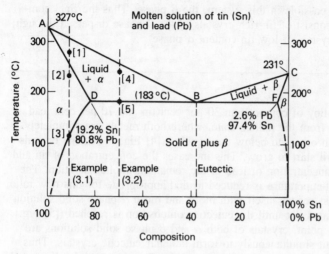

Fig. 3.14 Tin-lead phase equilibrium diagram

(b) For hypo-eutectic alloys, the solidus is the line ADB. Between the liquidus and the solidus the hypo-eutectic alloys will consist of the liquid solution of tin and lead plus crystals of the solid solution of α composition.

(c) Below the eutectic temperature the line separating the α phase from the α + β phase is called the *solvus* (see Fig. 3.13).

(d) For hyper-eutectic alloys, the solidus is the line BFC. Between the liquidus and the solidus, the hyper-eutectic alloys will consist of the liquid solution of molten lead and tin plus crystals of the β composition.

(e) Below the eutectic temperature, the line separating the α + β phase from the β phase is also called the solvus.

Example 3.1

Consider an alloy of composition 10 per cent tin and 90 per cent lead. During cooling from the molten state, where both metals are completely soluble in each other, to a temperature below the liquidus ([1] Fig. 3.14), crystals of the α phase solid solution start to grow. As in the previous diagrams solidification is complete when the solidus is reached. The solid alloy will consist of crystals of the α phase in solid solution ([2] Fig. 3.14). The composition of this solid solution will be 10 per cent tin in 90 per cent lead, as previously stated. However, as the temperature falls, it will eventually meet the solvus ([3] Fig. 3.14). At this point the solid solution will be *saturated* with tin. Further cooling to room (ambient) temperature will result in the tin precipitating out to form the only other solid solution possible in this system, the β phase. Thus the final composition will consist of tin-rich crystals of the β phase dispersed through a matrix of crystals of low tin content α phase.

Example 3.2
Consider an alloy of composition 30 per cent tin and 70 per cent lead. Upon cooling from the molten state, where both metals are completely soluble in each other, to below the liquidus ([4] Fig. 3.14), the crystals of α phase will start to grow. This increases the concentration of tin and reduces the concentration of lead in the remaining molten solution. The solidification temperature is reduced to that appropriate for this new ratio, and the process repeats itself with more and more α phase solid solution being precipitated out until the eutectic composition is reached ([5] Fig. 3.14). At this point, crystals of both α and β phase solid solutions are precipitated out simultaneously to form lamellar eutectic crystals. Thus the final composition will consist of crystals of α phase solid solution in a matrix of crystals of eutectic composition.

These examples explain the behaviour of the various types of soft solder in common use. *Tinman's solder* has a composition of 60 per cent tin and 40 per cent lead. Since this is approximately the eutectic composition this solder has the lowest melting point and also solidifies instantly with no 'pasty' range. These factors together with its relatively high tin content and low electrical resistance accounts for its widespread use for soldered joints in the electronics industry. On the other hand, a plumber requires a solder with a long pasty range which will solidify slowly and enable a wiped joint to be made. *Plumber's solder* has a composition of 80 per cent lead and 20 per cent tin so that there is a large temperature range between the liquidus and the solidus. At the same time the liquidus temperature is safely below that for pure lead so that there is no danger of melting the lead pipes or components being joined.

There are many other examples of binary alloys which could be quoted, but the three examples considered cover the three most common types of phase equilibrium diagrams.

3.11 Coring

So far, cooling above the liquidus has been assumed to be so slow that equilibrium is achieved as each change occurs. This rarely occurs in practice. One example, the cooling of a copper-nickel alloy under equilibrium conditions, will be considered in greater detail and the

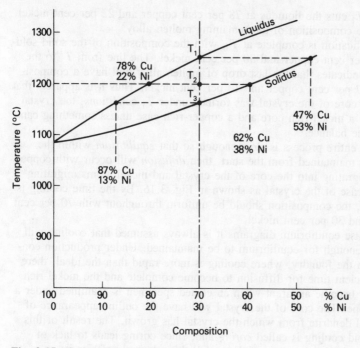

Fig. 3.15 Copper-rich Cu-Ni alloys

mechanism of solidification will then be compared with the same alloy cooled under production conditions.

Figure 3.15 shows part of the copper-nickel phase equilibrium diagram enlarged for clarity. It is convenient to consider an alloy of 70 per cent copper and 30 per cent nickel since solidification will conveniently centre on 1200°C. When the molten alloy cools to the liquidus small dendrites of copper-nickel solid solution commence to form. If a line is drawn from T_1 on the liquidus parallel to the composition axis until it cuts the solidus, it is apparent that the composition of the solid solution will be 47 per cent copper and 53 per cent nickel. Since the overall composition of the alloy is still 70 per cent copper and 30 per cent nickel, the fact that the newly formed dendrites have 53 per cent nickel will result in the remaining molten solution having less than 30 per cent nickel.

As the alloy cools down to 1200°C, the dendrites grow in size. A line drawn through T_2 parallel to the composition axis until it cuts the solidus indicates that the composition of the solid solution for this temperature will be 62 per cent copper and 38 per cent nickel. Thus between T_1 and T_2 the percentage of copper present in the dendrite has increased, whilst the percentage of nickel present in the dendrite has fallen. Since the line through T_2 cuts the liquidus at 78 per cent copper and 22 per cent nickel, this is the composition of the remaining molten alloy.

Solidification is complete at T_3, with the composition of the solid solution 70 per cent copper and 30 per cent nickel. The line from T_3 to the liquidus indicates that the last drop of molten alloy will have a composition of 87 per cent copper and 13 per cent nickel. Thus it is apparent that since the core of the crystal was formed under T_1 conditions, the crystal will have a nickel-rich core and a copper-rich case unless something can restore the balance.

If the entire process is slow enough so that *equilibrium* within the crystal is maintained from the start, then *diffusion* will occur with copper atoms migrating into the core of the crystal and nickel atoms migrating into the case of the crystal as shown in Fig. 3.16. By the time cooling is complete, the composition should be uniform throughout with 70 per cent copper and 30 per cent nickel.

In phase equilibrium diagrams it is always assumed that cooling will be slow enough for equilibrium to be maintained. Under production conditions in the foundry, where cooling is more rapid than the ideal, there is insufficient time for diffusion to become complete and the nickel-rich core will become apparent when an etched specimen is examined under a microscope. The core of the crystal will have the outline appearance of the initial dendrite from which the crystal has grown. The result of this more rapid cooling is called *coring* and, since coring leads to lack of uniformity in the structure of the metal, this adversely affects its mechanical properties. Coring can largely be eliminated by heat treat-

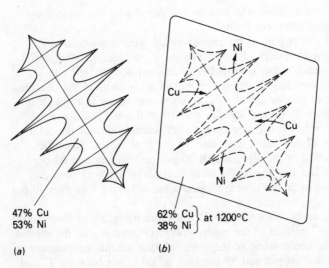

47% Cu
53% Ni

62% Cu
38% Ni } at 1200°C

(a)   (b)

Fig. 3.16 Diffusion during crystal growth (a) Dendritic nucleus at liquidus temperature (b) Diffusion of copper and nickel as crystal commences to grow

ment. The casting is heated to just below the solidus for the alloy concerned until diffusion is complete. Once diffusion is complete the rate of cooling is irrelevant. However, over-fast cooling creates stresses in the metal, whilst excessively long periods of heating and excessively slow cooling results in grain growth which may improve ductility but reduces mechanical strength and may cause machining problems.

4 Plain carbon steels

4.1 Ferrous metals

Ferrous metals and alloys are based upon the metallic element *iron*. The name ferrous comes from the Latin name for iron which is *ferrum*. Iron is a soft, grey metal and it is rarely found in the pure state outside the laboratory. The engineer usually finds it associated with the non-metal *carbon*, with which it forms solid solutions and the compound iron carbide. The carbon content is carried over from the smelting process during which the iron is extracted from its ore.

Since all the ferrous materials used by engineers contain iron in association with carbon, it could be argued that all such materials are ferrous alloys. However this term (ferrous alloy) is reserved for those ferrous materials containing additional metallic alloying elements in sufficient quantities to modify substantially the properties of the material, for example nickel-chrome alloy steels, or chrome-vanadium alloy steels, of great strength. Those 'alloys' containing only carbon as the main alloying element are referred to as wrought iron, plain carbon steels and plain cast irons. Since it is no longer widely used for engineering purposes, wrought iron will not be considered further in this text. Table 4.1 shows the relationship between the amount of carbon present and the resulting ferrous metal. It also gives some typical applications of those metals. These relationships will be considered further in Sections 4.4 and 4.5. The plain carbon steels will be considered in this chapter and the cast irons will be considered in Chapter 5 of this text. Alloy steels will be considered in *Engineering Materials: vol.2*. Wrought iron will not be considered in this text since it is no longer widely used for engineering purposes.

Table 4.1 Ferrous metals

Name	Group	Carbon content %	Some uses
Low carbon steel	Plain carbon steel	0.1 to 0.15	Sheet for pressing out such shapes as motor car body panels. Thin wire, rod, and drawn tubes.
	Plain carbon steel	0.15 to 0.3	General purpose workshop bars, boiler plate, girders
Medium carbon steel	Plain carbon steel	0.3 to 0.5	Crankshaft forgings, axles
		0.5 to 0.8	Leaf springs, cold chisels
High carbon steel	Plain carbon steel	0.8 to 1.0	Coil springs, wood chisels
		1.0 to 1.2	Files, drills, taps and dies
		1.2 to 1.4	Fine-edged tools (knives, etc.)
Grey cast iron	Cast iron	3.2 to 3.5	Machine castings

4.2 The iron-carbon system

Figure 4.1 shows the iron-carbon phase equilibrium diagram. Strictly, it should be called the iron-iron carbide diagram but conventionally it is called the iron-carbon diagram and this latter name will be used in this text. Comparison with the diagrams shown in Chapter 3 shows it to be of the combination type where two substances are completely soluble in each other in the liquid (molten) state but only partially soluble in each other in the solid state. Figure 4.1 is different from that considered in Chapter 3 because of the structural changes resulting from iron being allotropic, that is, it can exist in more than one form. These structural changes take place at 910°C and 1400°C.

(a) Below 910°C the iron forms body-centred cubic crystals.
(b) From 910°C to 1400°C it forms face-centred cubic crystals.
(c) Above 1400°C it reverts to body-centred cubic crystals.

These changes in lattice structure are accompanied by changes in volume as the atoms in the crystal lattice re-arrange themselves. For example, when iron is heated, it expands uniformly with temperature until it reaches 910°C whereupon it contracts slightly as the atoms re-arrange themselves into a more compact lattice, after which the material continues to expand uniformly again as shown in Fig. 4.2(a). A simple apparatus for demonstrating this phenomenon is shown in Fig. 4.2(b).

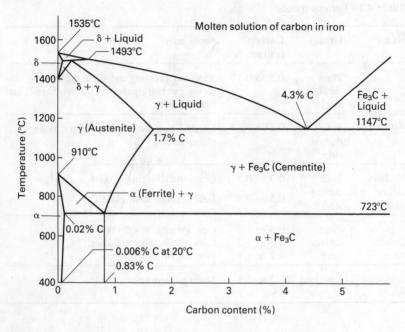

Fig. 4.1 Iron-carbon phase equilibrium diagram

These structural changes are accompanied by latent heat energy being taken in or given out. If an iron rod is cooled from above 910°C in a darkened room, it will suddenly glow with increased brightness. The change from face-centred to body-centred crystals releases latent heat energy more rapidly than it can be dissipated and the temperature of the rod momentarily rises and, for a moment, it glows more brightly. This phenomenon is call *recalescence*.

The iron-carbon phase equilibrium diagram appears to be very complex compared with those considered in Chapter 3. Fortunately this chapter is only concerned with the solid phases of the diagram, conventionally known as the 'steel section' of the full phase equilibrium diagram. The 'steel section' of the phase equilibrium diagram has been redrawn to larger scale in Fig. 4.3. It can be seen from Fig. 4.3 that there are only three important phases:

(a) *Ferrite* (α *phase*). This is a weak solution of carbon in body-centred cubic crystals of iron. There is a maximum of 0.03 per cent carbon in solid solution at 723°C, falling to 0.006 per cent carbon in solid solution at room temperature. (For all practical purposes it may be considered as 'pure' iron.) Ferrite is very soft, ductile and of relatively low strength.

(b) *Austenite* (γ *phase*). This is a much more concentrated solid solution

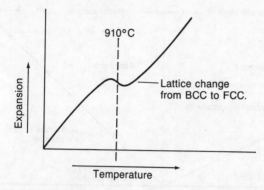

(a) Change in volume as crystal lattices rearrange themselves.

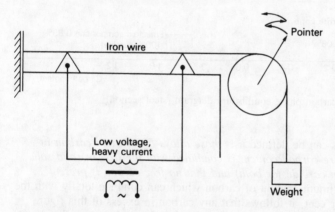

(b) Method of demonstrating volume changes with temperature.

Fig. 4.2 Effect of lattice change on volume

of carbon in iron than ferrite. Austenite is formed when carbon dissolves in face-centred cubic crystals of iron in the solid state. The maximum amount of carbon which can be held in solution with iron in the solid state is 1.7 per cent at 1150°C, as shown in Fig. 4.1. Although this is the upper limit of carbon which can be present in plain carbon steels, for all practical purposes there is no advantage in increasing the carbon content beyond about 1.2 to 1.4 per cent.

(c) *Cementite (iron-carbide phase).* An excess of carbon (C) combines with iron (Fe) to form iron carbide (Fe_3C). Each molecule of iron carbide contains three atoms of iron chemically combined with one atom of carbon. This is true up to the limit of 1.7 per cent carbon at room temperature, beyond which the excess carbon is precipitated out as 'free' or uncombined flakes of graphite.

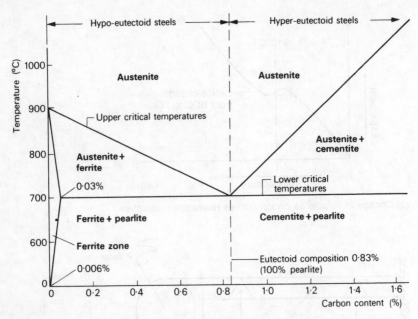

Fig. 4.3 Iron-carbon phase equilibrium diagram (steel section)

Thus steels can be defined as: '*those alloys of iron and carbon in which the entire carbon content is combined with the iron in solid solution or as iron carbide (or both) and that no free carbon is present*'. Since the maximum amount of carbon which can combine totally with the iron is 1.7 per cent, it follows that any carbon in excess of this figure will precipitate out as graphite flakes (graphite is an allotrope of carbon) and the resulting material will not be a steel but a cast iron, (see Chapter 6).

The steel section of the iron-carbon phase equilibrium diagram is very similar to the combination type of equilibrium diagram shown in Chapter 3. In the combination type diagram there was one eutectic composition at which both alloying elements crystallised out simultaneously at the same temperature to form a lamellar structure. However, in the steel section of the iron-carbon phase equilibrium diagram such transformations occur in the solid state and the point at which ferrite and cementite (iron carbide) precipitate out from the solid solution austenite is called the *eutectoid* point. This point occurs at a temperature of 723°C when 0.83 per cent carbon is present. The crystals which are precipitated out have a lamellar structure consisting of alternate layers of ferrite and cementite. This lamellar structure of ferrite and cementite is referred to as *pearlite* and is the toughest structure which can exist in plain carbon steels. When the carbon content is 0.83 per cent the steel consists entirely of pearlite and it has maximum toughness. Steels with a carbon content below 0.83 per cent are called *hypo-eutectoid* steels, whereas steels with a carbon content above 0.83 per cent are called *hyper-eutectoid* steels.

The transformations which occur during the cooling of a eutectoid composition (0.83 per cent carbon) steel are shown in Fig. 4.4(*a*). The

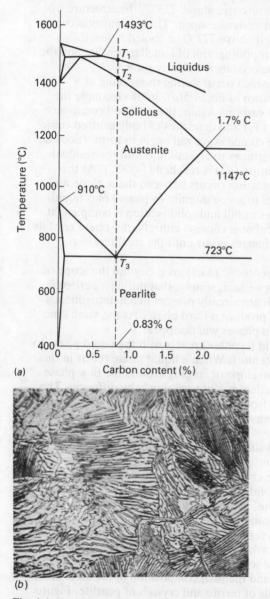

(*a*)

(*b*)

Fig. 4.4 Transformations for a 0.83 % carbon steel (*a*) Transformations at the eutectoid composition (*b*) Lamellar pearlite (× 600)

steel commences to solidify directly into austenite at the liquidus (T_1) and solidification is complete when the solidus is reached at temperature (T_2). The steel now consists entirely of γ phase crystals of austenite. At 723°C (T_3) the austenite suddenly changes into pearlite as shown. It remains as pearlite at all temperatures below 723°C. Should the temperature rise above 723°C, the structure will return to the solid solution austenite again. These changes occur each time the steel is heated above 723°C or cooled below it. Figure 4.4(b) shows a microphotograph of lamellar pearlite and the individual layers within the crystals are clearly shown.

The transformations which occur during the cooling of a hypoeutectoid steel are shown in Fig. 4.5(a). In this example the steel contains 0.5 per cent carbon. Again, the steel will commence to solidify at temperature (T_1) and dendrites of body centred cubic (BCC) crystals of α phase composition will begin to form. (See also: Fig. 4.1). This α phase continues to crystallise out of the residual molten steel until temperature (T_2) is reached (1493°C). At this temperature a peritectic reaction occurs between the α phase and the remaining molten steel to give austenite (γ phase) plus liquid. The temperature continues to fall and solidification is complete at temperature (T_3). The steel now consists entirely of γ phase crystals of austenite. No further changes occur until the steel reaches temperature (T_4).

A detailed study of peritectic reactions is beyond the scope of this book. However it can be said, simply, that during a peritectic reaction two phases which are already present in a heterogeneous mixture react together to produce a third phase. At the same time one or both of the original phases will disappear.

In this example a solid (α phase) reacts with the liquid phase to produce the γ phase (austenite). When a liquid phase reacts in this way it usually forms an envelope or coating around the new phase. This prevents further reaction except very slowly by diffusion. The term **peritectic** is derived from the Greek word "peri" which means around and refers to the envelope or coating which forms around the newly created phase.

When the steel cools slowly below (T_4) crystals of ferrite will start to grow in the austenite so that both α and γ phase crystals will be present. Since α phase crystals of ferrite contain rather less than 0.03 per cent carbon in solid solution, the carbon content of the remaining phase, austenite, will increase progressively as more and more ferrite is formed, until at 723°C (the eutectoid temperature) the structure will contain ferrite (> 0.03 per cent carbon) and austenite (0.83 per cent carbon) which is the euctectoid composition. Thus at (T_5) the austenite will change suddenly into the eutectoid composition of pearlite, and the final composition of the steel below (T_5) will consist of crystals of ferrite and crystals of pearlite. Figure 4.5(b) shows a typical microstructure for an annealed 0.5 per cent carbon steel.

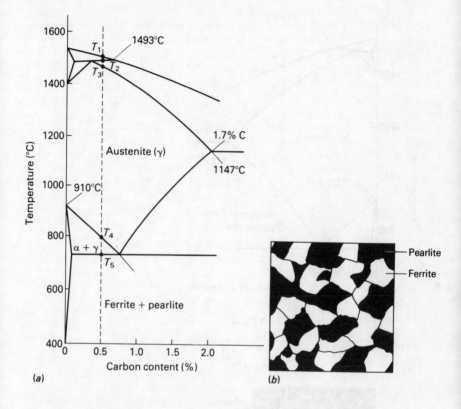

Fig. 4.5 Transformations for a 0.5% carbon steel (*a*) Transformations for hypo-eutectoid steels (*b*) Typical microstructure for a 0.5% carbon steel. (Annealed)

4.3 Critical change points

The construction of phase equilibrium diagrams from a family of cooling curves was explained in Chapter 3. The iron-carbon phase equilibrium diagram is also constructed from just such a family of cooling curves by connecting its *critical change points*. The change points are often referred to, simplistically, as the upper critical temperature (U.C.T.) and the lower critical temperature (L.C.T.). The critical change points, where changes in composition and structure occur, are also called *arrest* points since the time-temperature heating or cooling curve stops at these points as shown in Fig. 4.7, as the latent heat energy associated with change is taken in during heating or given out during cooling.

A_1 is the temperature at which the eutectoid transformations take place; that is the transformation of austenite into pearlite on cooling and

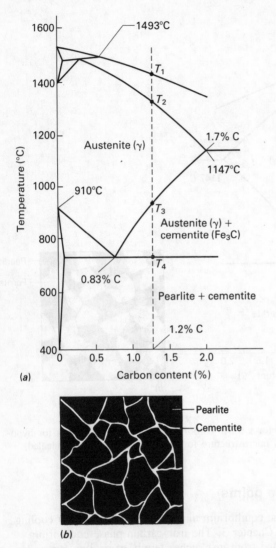

(a)

(b)

Fig. 4.6 Transformations for a 1.5% carbon steel (a) Transformations for hyper-eutectoid steels (b) Typical microstructure for a 1.5% carbon steel. (Annealed)

vice versa on heating. For plain carbon steels, A_1 is always constant and is 723°C.

A_3 is the temperature above which hypo-eutectoid steels are wholly austenitic (γ phase).

A_{cm} is the temperature above which hyper-eutectoid steels are wholly austenitic (γ phase).

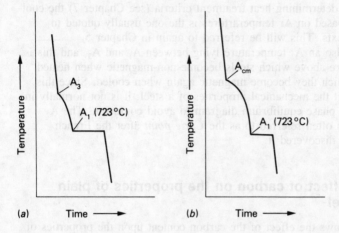

Fig. 4.7 Cooling curves for carbon steels (a) Cooling curve for a hypo-eutectoid steel (b) Cooling curve for a hyper-eutectoid steel

Due to what is known as thermal inertia the arrest points do not occur at exactly the same temperatures on heating curves as they do on cooling curves. Therefore the arrest points just described require further identification to indicate whether they were derived by heating or by cooling. This further notation makes use of the French word for heating which is *chauffage* and the French word for cooling which is *refroidissement*. Thus the critical change points on a time-temperature *heating* curve are called. Ac_1, Ac_3 and Ac_{cm}. Similarly the critical change points on a time-temperature *cooling* curve are called Ar_1, Ar_3 and Ar_{cm}. Their disposition on the phase equilibrium diagram for plain carbon steels is shown in Fig. 4.8. Since phase equilibrium diagrams are generally only used by

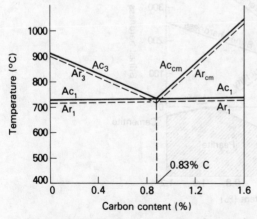

Fig. 4.8 Critical change points for carbon steels

engineers for determining heat treatment criteria (see Chapter 7) the cooling diagram based on Ar temperartures is the one usually quoted in engineering texts. This will be referred to again in Chapter 5.

There is also an A_2 temperature lying between A_1 and A_3, and this is the temperature above which steels become non-magnetic when heated and below which they become magnetic again when cooled. Since this does not affect the mechanical properties of a steel, it is not normally included on the phase equilibrium diagram to avoid confusion. The A_2 temperature is often referred to as the *Curie point* after the French physicist who discovered it.

4.4 The effect of carbon on the properties of plain carbon steel

Figure 4.9 shows the effect of the carbon content upon the properties of plain carbon steels which have been cooled slowly enough to enable them to achieve phase equilibrium. It can be seen from Fig. 4.9 that low carbon steels, consisting mainly of ferrite, are soft and ductile and relatively weak, reflecting the properties of the ferrite itself.

The increased amount of carbon in medium carbon steels promotes the formation of cementite. This results in an increased presence of pearlite, making such steels stronger, tougher and harder, but not so ductile.

When the carbon content reaches approximately 0.83 per cent the steel consists entirely of pearlite. This is the eutectoid composition and it produces plain carbon steel of maximum toughness and strength.

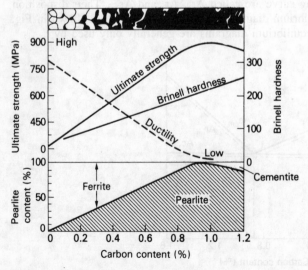

Fig. 4.9 Properties of plain carbon steels

Increasing the carbon content still further increases the amount of cementite (iron-carbide) present in the steel. Since the maximum amount of combined cementite occurred at 0.83 per cent carbon content, the formation of further cementite results in it appearing around the crystal boundaries. This increases the hardness and wear resistance of the steel, but at the expense of still further reduced strength and ductility.

4.5 Plain carbon steels

These are ferrous materials containing between 0.1 and 1.7 per cent carbon as the main alloying element. In addition, plain carbon steels contain the following elements by accident or by design.

Manganese: up to 1.5%
Phosphorus: up to 0.05%
Silicon: up to 0.3%
Sulphur: up to 0.05%

Manganese

Manganese is an essential constituent element since it ensures a sound ingot free from blow holes. Further, it combines with any sulphur present which would otherwise weaken the steel. In general, manganese raises the yield point (see Section 11.3) and increases the strength and toughness of the steel. However, it also increases the tendency of the steel to crack and distort when quench hardened (see Section 5.7) and, for this reason, the content should be kept below 0.5% in medium and high carbon steels.

Phosphorus

Phosphorus is an impurity carried over from the iron ore. It forms compounds which make the steel brittle and, therefore, should be removed as far as possible during the refinement processes. It should not be present in excess of 0.05%.

Silicon

Silicon is an impurity carried over from the iron ore. Its presence should be limited to between 0.1 and 0.3 per cent in the steels otherwise it can cause breakdown of the cementite which would result in weakness. Silicon has little direct effect upon the mechanical properties of plain carbon steels providing the amount present is limited to the percentage quoted above. It is often added to cast irons to prevent chill hardening (see Section 6.2).

Sulphur

Sulphur is an impurity carried over from the coke used in the blast furnace to extract the iron from its ore. Sulphur tends to combine with the iron to form iron sulphide which greatly weakens the steel. For this reason the sulphur content must be kept below 0.05 per cent and there should always be at least 5 times as much manganese present as there is sulphur. Fortunately, sulphur has a greater affinity for manganese than it has for steel and will combine with the manganese in preference to the iron. Unlike iron sulphide which weakens the steel, manganese sulphide has no such effect. Some free-cutting steels contain up to 0.2 per cent sulphur to improve their machineability at the expense of strength for lightly stressed turned parts. In this instance, the high sulphur content provides an extreme pressure lubrication between the chip and the tool and also causes the chips (swarf) to break up into easily disposable particles.

Table 4.2 gives the composition, properties and typical applications of some general purpose plain carbon steels, together with reference to their British Standard Specification. In comparing the properties of these steels it should be noted that some have had their properties modified by heat treatment. The state of the steel is, therefore, also stated in the table. Although various specifications have been quoted in Table 4.2, the most widely used is BS 970. This specification is concerned with wrought steels and is published in six parts:

BS 970: Part 1: 1972   covers carbon and carbon-manganese steels including free-cutting steels.
BS 970: Part 2: 1970   covers direct hardening alloy steels, including nitriding steels.
BS 970: Part 3: 1971   covers steels for case hardening.
BS 970: Part 4: 1970   covers stainless, heat-resisting and valve steels.
BS 970: Part 5: 1972   covers spring steels for hot-coiled springs.
BS 970: Part 6: 1973   covers SI metric values for use with BS 970 Parts 1 to 5 inclusive.

This text is only concerned with plain carbon steels as listed in Table 4.2 and as specified in BS 970: Part 1. The alloy steels covered in BS 970: Parts 2 to 5 inclusive will be considered in *Engineering Materials: Vol 2*.

Until 1970, steels in the United Kingdom were specified by a system of 'EN' numbers. These were a simple numerical listing and gave no indication of the properties and composition of the steel. Between 1970 and 1972 the British Standards Institution revised BS 970: Wrought Steels into a more logical and informative coding system using six symbols for each grade of steel. This code is built up as follows.

(a) The first three symbols are a number code indicating the type of steel.
  000 to 199 Carbon and carbon-manganese steels. The numbers indicate the manganese content × 100.
  200 to 240 Free-cutting steels. The second and third numbers indicate the sulphur content × 100.
  250   Silicon-manganese valve steels.

300 to 499 Stainless and heat-resisting steels.

500 to 999 Alloy steels.

(b) The fourth symbol is a letter code:

A —   The steel is supplied to a chemical composition determined by chemical analysis of a batch sample.

H —   The steel is supplied to a hardenability specification.

M —   The steel is supplied to a mechanical property specification.

S —   The material is a stainless steel.

(c) The fifth and sixth symbols are a number code indicating the mean carbon content. The actual mean carbon content × 100 is the code. Thus a steel of specification BS970.040A10 would be interpreted as follows:

BS970   indicates the specification;

040   classifies the steel as 'plain carbon', and indicates that the manganese content is 040/100 = 0.4 per cent.

A —   indicates that the composition has been determined by batch analysis.

10   indicates that the carbon content is 10/100 = 0.1 per cent.

Some examples of the use of the six-symbol code applied to plain carbon steels are given in Table 4.3.

The final factor to be considered in the coding of wrought steels is the *limiting ruling section*. This will be considered in greater detail in Sections 5.10 and 5.11 but, briefly, it is the maximum diameter bar of given composition which, after appropriate heat treatment, will attain its specified mechanical properties. For example, a plain carbon steel bar of composition 070M55 can attain condition 'R' (Table 4.4) after heat treatment providing it is not greater than 100 mm diameter. However, if it is to attain condition 'S' (Table 4.4), its diameter must be limited to 63 mm. Thus, in the first instance, the limiting ruling section is 100 mm diameter, and, in the second instance, the limiting ruling section is reduced to 63 mm diameter.

Table 4.2 Some plain carbons steels

Type of Steel	British Standard Specifications	Composition C%	Mn%	Condition	Properties Y.P. (MPa)	U.T.D. (MPa)	Elong'n (%)	Impact (J)	Hardness (H_B)	Applications
Low carbon steel	BS970.040A10	0.10	0.40	Process annealed after cold rolling	—	300	28	—	—	Car body panels produced by drawing and pressing.
	BS 15	0.20	—	As rolled	240	450	25	—	—	General purpose: mild steel. Welding quality, high tensile mild steel for building construction, etc.
	BS968	0.20	1.50	As rolled	350	525	20	—	—	
Casting steel BS 1504/161B		0.30	—	Annealed after casting to refine grain.	265	500	18	20	150	General purpose, medium strength castings for machining.
Medium carbon steel	BS 970.080M40	0.40	0.80	Toughened by quenching from 850°C, temper at 600°C	500	700	20	55	200	Axles, crankshafts, etc., under moderate stress.
	BS 970.070M55	0.55	0.70	Harden by quenching from 825°C. Temper at 600°C	550	750	14	—	—	Gears and machine parts subject to wear.

Table 4.2 continued Some plain carbon steels

Type of Steel	British Standard Specifications	Composition		Properties					Applications
		C%	Mn%		Y.P. (MPa)	U.T.S. (MPa)	Elong[n] (%)	Impact (J)	Hardness (H_B)	

Type of Steel	British Standard Specifications	C%	Mn%	Heat treatment	Y.P. (MPa)	U.T.S. (MPa)	Elong[n] (%)	Impact (J)	Hardness (H_B)	Applications
	—	0.70	0.35	Quench harden from 790/810°C in water. Temper at 150 to 300°C as appropriate.	—	—	—	—	780	Hand chisels, cold sets, screwdriver blades, blacksmith's tools, etc.
	BS4659: BW18	1.00	0.35	Quench harden from 760/780°C in water. Temper at 150 to 300°C as appropriate.	—	—	—	—	800	Taps, screwing dies, wood drills, press tools, hand (fitting) tools, files, measuring and marking out in instruments, etc.
	BS 4659: BW1C	1.20	0.35	Quench harden from 760/780°C in water or oil. Temper at 150 to 300°C as appropriate.	—	—	—	—	820	Fine edge tools, knives, files, surgical instruments.

Low Carbon Steels   High Carbon Steels

Table 4.3 Applications of the six symbol code

BS 970 spec	Description
070M26	A plain carbon steel with a composition of 0.26% carbon and 0.70% maganese. The letter 'M' indicates that the steel has to meet a prescribed mechanical property specification
150M36	As above except that the composition for this steel is: 0.36% carbon, 1.5% manganese
220M07	A low carbon free-cutting steel with a composition of 0.07% carbon and 0.20% sulphur. Again the letter 'M' indicates that the steel has to meet a prescribed mechanical property specification
070A20	A low carbon steel with a composition of 0.20% carbon and 0.70% manganese. The letter 'A' indicates that the steel has to meet a prescribed chemical composition specification

Table 4.4 Carbon and carbon manganese steels: derived from BS 970: Pt 1

Heat treatment condition symbol	Tensile strength range R_m (MPa)	Brinell hardness number range – H_B
P	550–700	152–207
Q	620–770	179–229
R	690–850	201–255
S	770–930	223–277
T	850–1000	248–302

Steel	P LRS	P R_e	P A	P I	P $R_{p0.2}$	Q LRS	Q R_e	Q A	Q I	Q $R_{p0.2}$	R LRS	R R_e	R A	R I	R $R_{p0.2}$	S LRS	S R_e	S A	S I	S $R_{p0.2}$	T LRS	T R_e	T A	T I	T $R_{p0.2}$
070M20	19	355	20	41	340	—	—	—	—	—	—	—	—	—	—	—	—	—	—	—	—	—	—	—	—
070M26	29	355	20	41	325	—	—	—	—	—	—	—	—	—	—	—	—	—	—	—	—	—	—	—	—
080M30	63	340	18	34	310	—	—	—	—	—	—	—	—	—	—	—	—	—	—	—	—	—	—	—	—
080M36	—	—	—	—	—	13	415	16	34	400	13	465	16	34	450	—	—	—	—	—	—	—	—	—	—
080M40	—	—	—	—	—	19	415	16	34	400	19	465	16	34	450	—	—	—	—	—	—	—	—	—	—
080M46	—	—	—	—	—	29	400	16	34	370	29	450	16	—	415	—	—	—	—	—	—	—	—	—	—
080M50	—	—	—	—	—	63	385	16	—	355	63	430	16	—	400	13	525	14	—	510	—	—	—	—	—
070M55	—	—	—	—	—	100	370	16	—	340	100	415	14	—	385	29	495	14	—	465	—	—	—	—	—
120M19	100	355	18	28	325	29	450	16	47	415	19	510	16	34	495	—	—	—	—	—	—	—	—	—	—
150M19	150	340	18	27	310	63	430	16	54	415	29	510	16	41	480	63	480	14	—	450	—	—	—	—	—
120M28	—	—	—	—	—	100	415	16	41	385	29	510	16	34	480	—	—	—	—	—	—	—	—	—	—
150M28	—	—	—	—	—	150	400	18	47	370	63	480	16	41	450	13	510	16	34	555	13	570	12	—	555
150M36	—	—	—	—	—	150	415	18	41	385	29	510	16	34	480	19	570	14	34	555	19	570	12	—	555
216M28	63	355	20	34	325	150	400	18	47	370	63	480	16	41	450	29	555	14	41	525	—	—	—	—	—
212M36	100	340	20	34	310	19	430	18	41	415	13	495	16	54	480	—	—	—	—	—	—	—	—	—	—
225M36	—	—	—	—	—	63	400	18	34	370	29	480	16	34	450	—	—	—	—	—	—	—	—	—	—
216M36	100	340	20	34	310	63	400	18	34	370	29	480	18	34	450	13	540	14	27	525	13	635	12	34	620
212M44	—	—	—	—	—	63	400	18	34	370	63	465	16	34	430	—	—	—	—	—	—	—	—	—	—
225M44	—	—	—	—	—	100	400	—	—	—	100	450	16	34	415	29	525	14	27	495	13	600	12	27	585

LRS = Limiting ruling section, A = Elongation % $R_{p0.2}$ = 0.2% proof stress (MPa)
R_e = Yield stress, MPa I = Izod impact value (J)

5 Heat treatment of plain carbon steels

5.1 Heat treatment processes

Plain carbon steels and alloy steels are among the relatively few engineering materials which can be usefully heat-treated in order to vary their mechanical properties. This is because of the structural changes which can take place within solid iron-carbon alloys. The various heat treatment processes appropriate to plain carbon steels are:

(*a*) annealing;
(*b*) normalising;
(*c*) hardening;
(*d*) tempering;

In all the above processes the steel is heated slowly to the appropriate temperature for its carbon content and then cooled. It is the *rate of cooling* which determines the ultimate structure and properties which the steel will have, providing that the initial heating has been slow enough for the steel to have reached phase equilibrium at its process temperature. Before describing the processes listed above it is necessary to establish some basic principles.

Recrystallisation

During cold-working processes, the grain of the metal becomes distorted and internal stresses are introduced into the metal. If the temperature of the cold-worked metal is now raised sufficiently, nucleation occurs and 'seed' crystals form at the grain boundaries at points of maximum internal stress. The more severe the cold-working and the greater the internal stress, the lower will be the temperature at which nucleation occurs for a given metal. The principle of recrystallisation and nucleation is shown in Fig. 5.1. The minimum temperature at which the reformation of the crystals occurs is called the *temperature of recrystallisation*. At temperatures above the recrystallisation temperature, the kinetic energy of the atoms on the edges of the distorted grains increases. This allows these edge atoms to move away and attach themselves to the newly formed nuclei which will then begin to grow into crystals. This process continues until all the atoms of the original, distorted crystals have been transferred. Since, after severe cold-working, more nuclei form than the number of original grains, the grain structure after recrystallisation is usually finer than the original grain structure before cold-working. Thus a degree of grain refinement occurs.

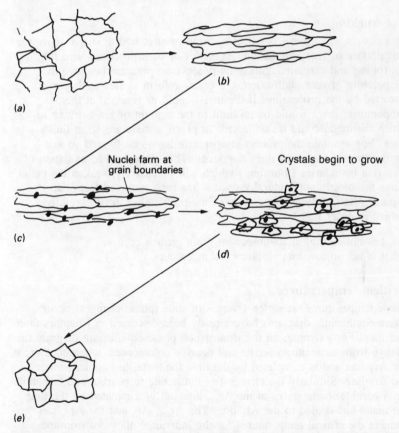

Fig. 5.1 Recrystallisation (*a*) Before working (*b*) After cold-working (*c*) Nucleation commences at recrystallisation temperature (*d*) Crystals commence to grow as atoms migrate from the original crystals and attach themselves to the nuclei (*e*) After annealing is complete the grain structure is restored

Cold-working

This occurs when metal is bent, squeezed or stretched to shape *below* the temperature of recrystallisation. Examples of such processes are: pressing out car body panels, cold-drawing rods, wires and tubes, cold-heading rivets, and cold-rolling strip and sheet metal. Cold-working results in distortion of the grain of the metal and, eventually, the metal becomes so stiff and brittle that it breaks. (This is what happens in a tensile test (see Section 11.2.) Metal which has become harder and stiffer as a result of cold-working is said to be *work-hardened*. The metal must not be allowed to become excessively work-hardened or it will be prone to fracture. Once work-hardened it needs to be annealed to restore its grain structure before further cold working is performed upon it. The heat treatment process of annealing is described in Section 5.2.

Hot-working

This occurs when metal is bent, squeezed or stretched to shape *above* the temperature of recrystallisation. Examples of such processes are: forging, hot-rolling and extrusion. Since the process temperature is above the temperature of recrystallisation, the grains reform as fast as they are distorted by the processing. If the metal could be retained at this temperature, there would be no limit to the amount of hot-working to which the metal could be subjected. In practice there are strict limitations. For example the process temperature has to be limited so that overheating of the metal does not occur. This could lead to oxidation of the grain boundaries ('burning') which will excessively weaken the metal. Since the metal cools naturally once it has been removed from the furnace, the finishing temperature of the process has to be carefully judged in order that:

(*a*) it is not so high that subsequent grain growth occurs;
(*b*) it is not so low that surface cracking occurs.

Critical temperatures

These temperatures, at which changes of state (phase changes) occur on phase equilibrium diagrams have already been discussed in Chapters three and four. For example, on the iron-carbon phase equilibrium diagram the change from austenite to ferrite and pearlite commences, when cooling, at the Ar_3 line and is completed by the time the metal has cooled slowly to the Ar_1 line. Similarly the change from austenite to pearlite and cementite (iron carbide) commences at the Ar_{cm} line and is completed by the time the metal has cooled to the Ar_1 line. The Ar_{cm}, Ar_3 and the Ar_1 lines connect the critical temperatures for the individual alloys of iron and carbon.

5.2 Annealing processes

All annealing processes are concerned with rendering steel soft and ductile so that it can be cold-worked and/or machined. There are three basic annealing processes, and these are:

(a) *stress-relief annealing* at sub-critical temperatures (also known as 'process annealing' and 'interstage annealing');
(b) *spheroidised annealing* at sub-critical temperatures;
(c) *full annealing* for forgings and castings.

The process chosen depends upon the carbon content of the steel, its pre-treatment processing, and its subsequent processing and use. Figure 5.2(a) shows the temperature bands for the annealing processes superimposed on the iron-carbon phase equilibrium diagram. In all annealing processes the cooling rate is as slow as possible.

5.3 Stress-relief annealing

This process is reserved for steels below 0.4 per cent carbon content. Such steels will not satisfactorily quench harden (Section 5.7) but, as they are relatively ductile, they are frequently cold-worked and become work-hardened. Since the grain structure will have been severely distorted by the cold-working, recrystallisation can commence at 500°C but, in practice, annealing is usually carried out between 630°C and 700°C to speed up the process and limit grain growth. The rate of cooling and the length of time for which the steel is heated depends upon the subsequent processing and use to which the material is going to be put. If further cold-working is to take place then increased ductility and malleability will be required. This is achieved by prolonging the heating and slowing the cooling to encourage grain growth. However, if grain refinement and strength and toughness is of more importance, then heating and cooling should be more rapid.

5.4 Spheroidising annealing

It has already been stated that crystals of pearlite have a laminated structure consisting of alternate layers of cementite and ferrite. When steels containing more than 0.4 per cent carbon are heated to just below the critical temperature (650°C to 700°C) the cementite in the crystals tends to 'ball up'. This is referred to as the aspheroidisation of pearlitic cementite and the process is shown diagrammatically in Figure 5.2(b). Since the temperatures involved are sub-critical, no phase changes take place and aspheroidisation of the cementite is purely a surface tension effect.

If the layers of cementite are relatively coarse prior to annealing, they take too long to break down and tend to form coarse globules (spheroids)

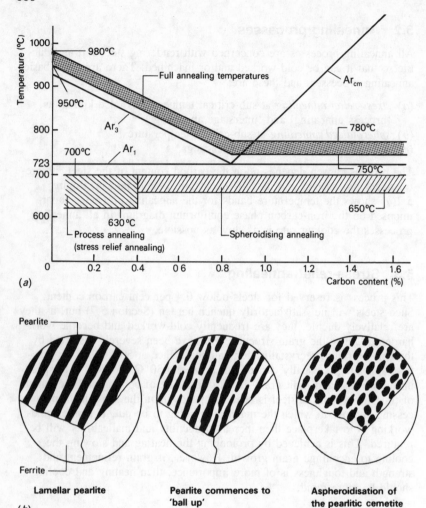

Fig. 5.2 Annealing (*a*) Annealing temperatures (*b*) Spheroidised annealing

of cementite. This, in turn, leads to impaired physical properties and machined surfaces with a poor finish. Thus, grain refinement by a quench treatment prior to aspheroidisation is recommended to produce fine globules of cementite. The process is most effective when it is used to soften plain carbon steels containing more than 0.4 per cent carbon and which have been either work-hardened or have been quench-hardened. After spheroidising annealing the steel can be cold-worked and it will machine freely to a good surface finish. Furthermore, steel which has been subjected to spheroidising annealing will re-harden more uniformly and with less chance of cracking. As with any other annealing process, slow cooling is required after the heating cycle. It is usual to turn off the furnace and allow the furnace and charge to cool down slowly together.

5.5 Full annealing

Plain carbon steels solidify at temperatures well above the temperatures with which heat treatment processes are concerned, and as a result large castings, well insulated by the sand mould, take a very long time to cool down. Similarly, large forgings, although hot-worked at temperatures well below their melting points are, nevertheless, processed at temperatures substantially above their upper critical temperatures for relatively long periods of time. In both cases grain growth is excessive and the physical properties of the metal are impaired. The ferrite settles out along the crystal boundaries of the coarse grains of austenite and also within the grains to provide a mesh-like structure, as shown in Fig. 5.3. This is called a *Widmanstätten structure*.

Reference back to Fig. 5.2(*a*) shows that to render the steel usable for certain applications it has to be reheated to approximately 50°C above the Ar_3 line for hypo-eutectoid steels and approximately 50°C above the Ar_1 line for hyper-eutectoid steel. This results in the formation of fine grains of austenite which transform into relatively fine grains of ferrite and pearlite or pearlite and cementite (depending upon the carbon content) as the steel is slowly cooled to room temperature, usually in the furnace.

5.6 Normalising

The normalising temperatures for plain carbon steel are shown in Fig. 5.4. The process resembles full annealing except that whilst in annealing

Fig. 5.3 Widmanstätten structure

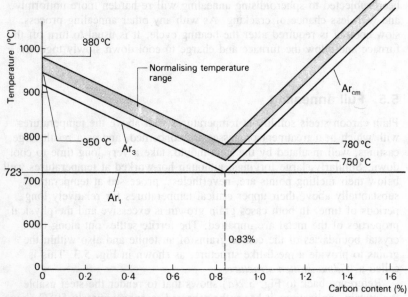

Fig. 5.4 Normalising temperatures for plain carbon steels

the cooling rate is deliberately retarded, in normalising the cooling rate is accelerated by taking the work from the furnace and allowing it to cool in free air. Provision must be made for the free circulation of cool air, but draughts must be avoided.

In the normalising process, as applied to hypo-eutectoid steels, it can be seen that the process temperature is the same as for full annealing. After 'soaking' the steel at the process temperature to ensure conversion to fine grain austenite, the more rapid cooling associated with the normalising process results in the transformation of the fine grain austenite into fine grain ferrite and pearlite. The relatively rapid cooling avoids the grain growth associated with annealing.

In the normalising process, as applied to hyper-eutectoid steels, it can be seen that the steel is heated to approximately 50°C above the Ar_{cm} line. This ensures that the transformation to fine grain austenite corrects any grain growth or grain distortion that may have occurred previously. Again, the steel is cooled in free air and the austenite transforms into fine grain pearlite and cementite. The fine grain structure resulting from the more rapid cooling associated with normalising gives improved strength and toughness to the steel but reduces its ductility and malleability. The increased hardness and reduced ductility allows a better surface finish to be achieved when machining. (The excessive softness and ductility of full annealing leads to local tearing of the machined surface.) However, the level of ductility and malleability achieved by normalising is not sufficient for more than limited cold-working. Normalising is frequently used for stress relieving between rough machining and the finish machining of large castings and forgings to avoid subsequent 'movement' due to the slow release of internal stresses and loss of accuracy. At one time large castings and forgings were left outside to 'weather' for up to a year or more after rough machining to ensure that the workpieces became stabilised. Although highly successful this procedure tied up an excessive amount of working capital and space and nowadays heat treatment is preferred as the work in progress is turned round more quickly.

5.7 Quench hardening

Figure 5.5 shows the temperature band from which plain carbon steels are cooled when they are quench hardened. It can be seen that this temperature band is the same as for full annealing. The band is not continued below 0.4 per cent carbon for, although some grain refinement and toughening occurs, no appreciable hardening takes place. If a plain carbon steel with a carbon content above 0.4 per cent is quenched (cooled very rapidly) from the appropriate temperature for its carbon content as shown in Fig. 5.5, there is insufficient time for the equilibrium transformations previously described to take place and the steel becomes appreciably harder. The final hardness will depend solely upon the carbon content and the rate of cooling.

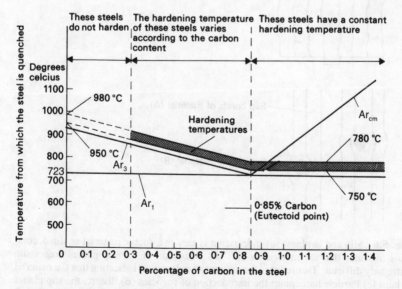

Fig. 5.5 Hardening temperatures for plain carbon steels

The reason for this increase in hardness can be briefly described by reference to Fig. 5.6. In the annealed condition metals can be formed by bending, stretching or squeezing them to shape. This is possible because the orderly arrangement of atoms in the crystals allows individual layers

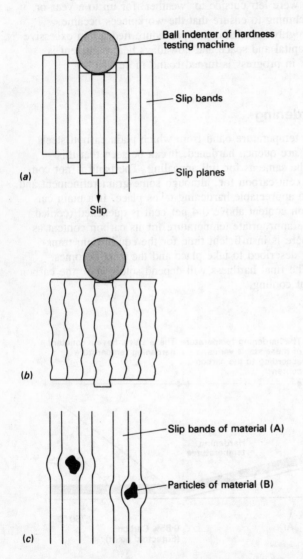

Fig. 5.6 Slip and hardness (*a*) Indentation is easy in a ductile material as slip occurs. This indicates that the material is soft (*b*) Distortion of the slip planes makes slip extremely difficult. This reduces the amount of indentation indicating that the material is hard (*c*) Particle hardening: the introduction of particles (B) distorts the slip planes and makes slip difficult

of atoms — called *slip planes* — to slide over each other as shown in Fig. 5.6(*a*). However if distortion of the lattice occurs as shown in Fig. 5.6(*b*) or particles of another material are introduced as shown in Fig. 5.6(*c*), then slip cannot occur so easily and the metal becomes hard and brittle.

When a steel is heated to its hardening temperature it becomes austenitic. If it is cooled quickly, the equilibrium transformations into pearlite and ferrite or pearlite and cementite do not have time to take place. Instead, a structure called *martensite* is formed. This is the hardest structure that it is possible to produce in a plain carbon steel and, under the microscope, it appears as acicular (needle-shaped) crystals as shown in Fig. 5.7. Actually these are sections through disc-shaped plates. What has happened is that the face-centred crystals of austenite have changed to body centred crystals below the Ar_1 line as usual but, because of the rapid cooling, there has not been time for the cementite to form and the body centred crystals are a super-saturated solid solution of carbon in iron (martensite). This so distorts the lattice structure that slip virtually becomes impossible and the steel becomes very hard and brittle.

Large components do not cool as quickly as small components and may not achieve the *critical cooling rate* necessary for maximum hardness. The critical cooling rate is defined as the slowest cooling rate (quenching rate) which will produce a martensitic structure throughout the mass of the steel. If this cooling rate is not achieved some pearlite will be formed and the steel will be tougher but substantially less hard. However, there is no virtue in exceeding the critical cooling rate to any great extent. Once maximum hardness has been achieved any increase in the cooling rate will only result in cracking and distortion of the workpiece. Further, there is no particular advantage in heating hyper-eutectoid steels above their Ar_{cm} temperature when hardening them and, in practice, the hardening temperature for hyper-eutectoid steels is just above the Ar_1 temperature (see Fig. 5.5). Quenching hyper-eutectoid steels from this lower temperature helps to prevent grain growth, cracking and distortion. The critical cooling rate can be substantially reduced by the addition of alloying elements to the steel. This enables thicker

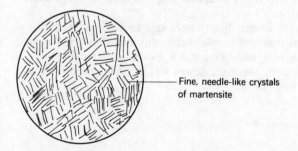

Fine, needle-like crystals of martensite

Fig. 5.7 Martensitic structure

components to be hardened with less chance of cracking and distortion and is one of the most important reasons for alloying steels. Alloy steels and their heat treatment will be dealt with in detail in *Engineering materials: vol 2*.

5.8 Quenching media

The most commonly used quenching media in increasing order of severity are:

(*a*) compressed air blast;
(*b*) oil;
(*c*) water;
(*d*) brine (10 per cent solution).

The choice of quenching bath depends upon the type of steel being treated and the resultant properties required. Brine, which is a solution of common salt and water is only occasionally used to provide very rapid cooling for plain carbon tool steels and case-hardening steels where maximum hardness is required. Such severe quenching can lead to cracking in all but the simplest components, and plain water and quenching oils are most commonly used for plain carbon and alloy steels. To avoid cracking and distortion the quenching rate should be no greater than that needed to give the required properties in the workpiece. Water provides a quenching rate approximately three times as great as oil and is usually used for plain carbon steel. Oil quenching is usually used with very high carbon steels (1.2 to 1.4 per cent) and alloy steels. Air blast quenching is usually reserved for high speed steel tools and components of small section. The alloy content of such steels is sufficiently high to reduce the critical cooling rate to a very low level.

As soon as the heated workpiece is plunged into the quenching bath it becomes surrounded by a blanket of vaporised quenching media. Since this vapour has a low thermal conductivity it slows the cooling process. Therefore the work must be constantly agitated in the quenching bath to disperse the vapour as it forms and keep the work in contact with the liquid. Agitation of the quenching bath also helps to keep its temperature uniform.

Care must be taken to ensure distortion is kept to a minimum during quenching, and for this reason it is essential to dip long thin components vertically into the quenching bath. Figure 5.8 shows how cracking and distortion can occur both by incorrect design and incorrect quenching and how these faults may be avoided.

5.9 Tempering

A fully hardened plain carbon steel is brittle and hardening stresses are present. In such a condition it is of little practical use and it has to be

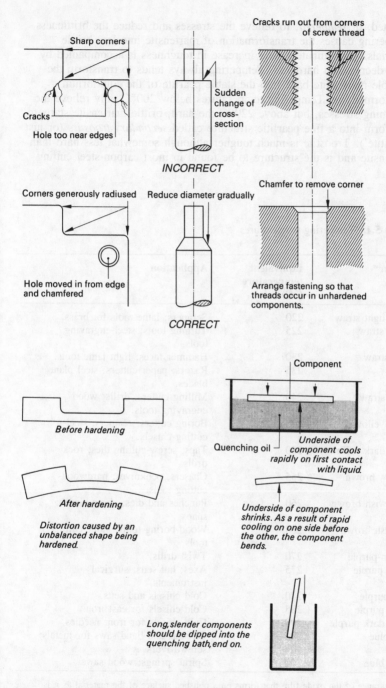

Sharp corners

Cracks

Hole too near edges

Cracks run out from corners of screw thread

Sudden change of cross-section

INCORRECT

Corners generously radiused

Reduce diameter gradually

Hole moved in from edge and chamfered

CORRECT

Chamfer to remove corner

Arrange fastening so that threads occur in unhardened components.

Before hardening

After hardening

Distortion caused by an unbalanced shape being hardened.

Component

Quenching oil

Underside of component cools rapidly on first contact with liquid.

Underside of component shrinks. As a result of rapid cooling on one side before the other, the component bends.

Long, slender components should be dipped into the quenching bath, end on.

Fig. 5.8 Causes of cracking and distortion

reheated, or *tempered*, to relieve the stresses and reduce the brittleness. Tempering causes the transformation of martensite into less brittle materials. Unfortunately, any increase in toughness is accompanied by some decrease in hardness. Tempering always tends to transform the unstable martensite back into the stable pearlite of the equilibrium transformations. Tempering temperatures below 200°C only relieve the hardening stresses, but above 220°C the hard, brittle martensite starts to transform into a fine pearlitic structure called *secondary troostite* (or just 'troostite'). Troostite is much tougher although somewhat less hard than martensite and is the structure to be found in most carbon-steel cutting tools.

Table 5.1 Tempering temperatures

Colour*	Equivalent temperature (°C)	Application
Very light straw	220	Scrapers; lathe tools for brass.
Light straw	225	Turning tools; steel-engraving tools.
Pale straw	230	Hammer faces; light lathe tools.
Straw	235	Razors; paper cutters; steel plane blades.
Dark straw	240	Milling cutters; drills; wood-engraving tools.
Dark yellow	245	Boring cutters; reamers; steel-cutting chisels.
Very dark yellow	250	Taps; screw-cutting dies; rock drills
Yellow-brown	255	Chasers; penknives; hardwood-cutting tools.
Yellowish brown	260	Punches and dies; shear blades; snaps.
Reddish brown	265	Wood-boring tools; stone-cutting tools.
Brown-purple	270	Twist drills.
Light purple	275	Axes; hot setts; surgical instruments.
Full purple	280	Cold chisels and setts.
Dark purple	285	Cold chisels for cast iron.
Very dark purple	290	Cold chisels for iron; needles.
Full blue	295	Circular and band saws for metals; screwdrivers.
Dark blue	300	Spiral springs; wood saws.

*Appearance of the oxide film that forms on a polished surface of the material as it is heated.

Tempering above 400°C causes any cementite particles present to 'ball-up' giving a structure called *sorbite*. This is tougher and more ductile than troostite and is the structure used in components subjected to shock loads and where a lower order of hardness can be tolerated. It is normal to quench the steel once the tempering temperature has been reached. Table 5.1 gives the tempering temperatures for various applications of plain carbon steel components.

5.10 Mass effect

It has already been stated that the hardness of a plain carbon steel depends upon its carbon content and the rate of cooling from the hardening temperature. Obviously a thin component will cool more quickly than a thick component if both are quenched from the same temperature in the same quenching bath.

In a thick component the heat will be trapped at the centre so that the core of the component cools more slowly than the outer layers. This leads to a variation in hardness across a section of the component as shown in Fig. 5.9(a). This variation in hardness is referred to as *mass effect*.

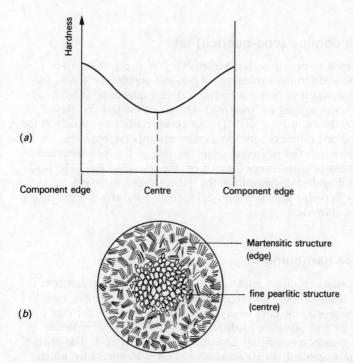

Martensitic structure (edge)

fine pearlitic structure (centre)

Fig. 5.9 Mass effect (hardenability)

Plain carbon steels have a highly critical cooling rate and, therefore, large sections cannot be fully hardened throughout as shown in Fig. 5.9(*b*). Thus plain carbon steels are said to have a *poor hardenability*. On the other hand a 3 per cent nickel steel containing only 0.3 per cent carbon will harden uniformly throughout its section because it has a relatively low critical cooling rate. Such a steel is said to have *good hardenability*. Therefore, hardenability can be defined as the ease with which hardness is obtained in a material.

Hardness and hardenability should not be confused. It has already been stated that a 1.0 per cent plain carbon steel has poor hardenability compared with a 3 per cent nickel steel containing only 0.3 per cent carbon. However, because of its higher carbon content the 1.0 per cent plain carbon steel will have a very much higher surface hardness.

Lack of uniformity of structure and hardness in steels with poor hardenability characteristics can seriously affect their mechanical properties. For this reason it is necessary to specify the maximum diameter or *limiting ruling section* (see Section 4.5) for which the stated mechanical properties can be achieved under normal heat treatment conditions. One of the main reasons for adding alloying elements such as nickel and chromium to steels is to reduce the mass effect and increase the ruling section for which prescribed properties can be achieved.

5.11 The Jominy (end-quench) test

This test is used to determine the hardenability of steels. It involves heating a specimen to the hardening temperature appropriate for its carbon content so that it is fully austenitic, and then quenching it by spraying a jet of water against its lower end. Details of the test and the specimen are shown in Fig. 5.10. The specimen cools very rapidly at the quenched end and progressively less rapidly towards the opposite (shouldered) end. A flat is ground along the side of the cold specimen and its hardness is tested every 3 mm from the quenched end. The hardness is plotted against distance from the quenched end to give a hardenability curve as shown in Fig. 5.11 which compares a plain carbon steel with an alloy steel.

5.12 Case hardening

Often components require a hard case to resist wear and a tough core to resist shock loads. These two properties do not exist in a single steel since, for toughness, the core should not exceed 0.3 to 0.4 per cent carbon, whilst, to give adequate hardness, the surface of the component should have a carbon content of approximately 1.0 per cent. The usual solution to this problem is *case-hardening*. This is a process by which carbon is added to the surface layers of a low carbon plain or low alloy

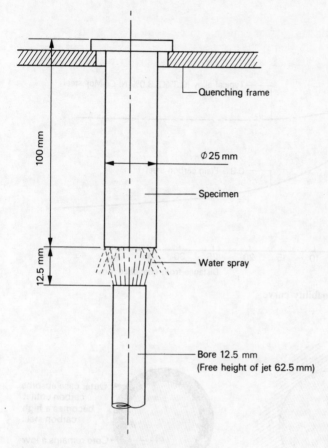

Fig. 5.10 Jominy end-quench test

steel component to a carefully regulated depth, after which the component goes through successive heat treatment processes to harden the case and refine the core. Thus the process has two distinct steps as shown in Fig. 5.12:

(*a*) *carburising* (the addition of carbon);
(*b*) *heat treatment* (hardening and core refinement).

Carburising makes use of the fact that low carbon steels (approximately 0.1 per cent carbon) absorb carbon when heated to the austenitic condition. Various carbonaceous materials are used in the carburising process.

(*a*) *Solid media* such as bone charcoal or charred leather, together with an energiser such as sodium and/or barium carbonate. The energiser makes up to 40 per cent of the total composition.

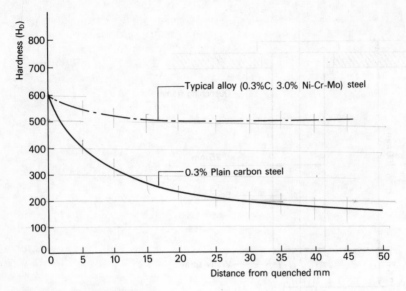

Fig. 5.11 Hardenability curve

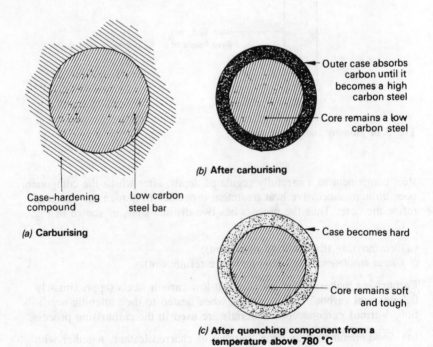

Fig. 5.12 Case hardening

(b) *Fused salts* such as sodium cyanide, together with sodium carbonate and varying amounts of sodium and barium chloride. Since cyanide is a deadly poison and represents from 20 per cent to 50 per cent of the total content of the molten salts, stringent safety precautions must be taken in its use. The components to be carburised are immersed in the molten salts.

(c) *Gaseous media* are increasingly used now that 'natural' gas (methane) is widely available. Methane is a hydrocarbon gas containing organic compounds of carbon which are readily absorbed into steel. The methane gas is often enriched by the vapours given off when oil is 'cracked' by heating it in contact with platinum which acts as a catalyst.

It is a fallacy to suppose that carburising hardens steel. Carburising merely adds carbon to the outer layers and leaves the steel in a fully annealed condition with a coarse grain structure. Therefore additional heat treatment processes are required to harden and refine the case and to refine and toughen the core. Reference to Fig. 5.13 will clarify the following descriptions of these heat treatment processes.

(a) *Refining the core.* Since the core has a content of less than 0.3 per cent carbon, the correct annealing temperature is approximately 870°C which is well below the carburising temperature of 950°C which caused the grain growth. After raising the component to 870°C it is water quenched to ensure a fine grain. Although the

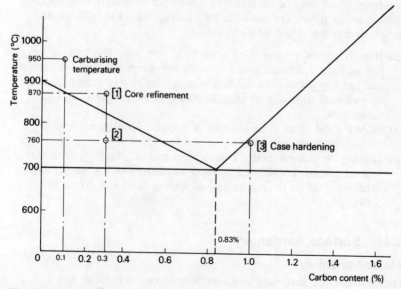

Fig. 5.13 Case-hardening temperatures

temperature of 870°C is correct for the low carbon core of the component (temperature [1] Fig. 5.13) it is excessively high for the high carbon case of the component (temperature [2] Fig. 5.13).

(b) *Refining and hardening the case.* Since the case has a carbon content of approximately 1.0 per cent carbon its correct hardening temperature is 760°C. Therefore the component is reheated to this temperature (temperature [3] Fig. 5.13) and again quenched. This hardens the case and ensures that it has a fine grain. The temperature of 760°C is too low to cause grain growth in the hypo-eutectic core providing the component is heated rapidly through the range 650°C to 760°C during the reheating and quenched without soaking at the hardening temperature.

(c) *Tempering.* It is advisable to relieve any quenching stresses present in the component by tempering it at about 200°C to 220°C.

The above heat treatment sequence is used to give ideal results in stressed components. However, in the interests of speed and economy the process is often simplified where components are lightly stressed or where alloy steels are used having less critical grain growth and quenching characteristics.

5.13 Localised case hardening

It is often not desirable to harden a component all over. For example it is undesirable to case-harden screw threads. Not only would they be extremely brittle, but any distortion occurring during carburising and hardening could only be corrected by expensive thread-grinding operations. Various means are available for avoiding the local infusion of carbon during the carburising process.

(a) Heavily copper plating those areas to be left soft. The layer of copper prevents intimate contact between the component and the carbon and prevents carburisation. Note that copper plating cannot be used for salt-bath treatment as cyanide dissolves the copper from the component.

(b) Encasing the areas to be left soft in fire clay. Mostly used when pack-carburising.

(c) Leaving on surplus metal. This is machined off, together with the infused carbon, between carburising and hardening. Although more expensive, this is the surest way of leaving local soft areas, (see Fig. 5.14).

5.14 Surface hardening

Flame hardening

Localised surface hardening can also be achieved in medium and high carbon steels and some cast irons by rapid local heating and quenching.

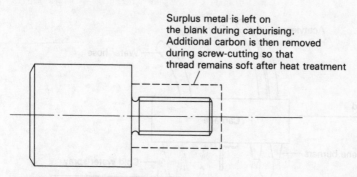

Fig. 5.14 Localised case-hardening

Figure 5.15(*a*) shows the principle of flame hardening. A carriage moves over the workpiece so that the surface is rapidly heated by an oxy-acetylene or an oxy-propane flame. The same carriage carries the water quenching spray. Thus the surface of the workpiece is heated and quenched before its core can rise to the hardening temperature. This process is often used for hardening the slideways of machine tools, for example lathe bed-ways.

Induction hardening

Figure 5.15(*b*) shows how the same surface hardening effect can be produced by high frequency electromagnetic induction. The induction coil surrounding the component is connected to a high frequency alternating current generator. This induces high frequency eddy currents in the component causing it to become hot. When the hardening temperature has been reached, the current is switched off and a water spray quenches the component. The induction coil can be made from copper tube which also carries the quenching water or oil. This technique is often used for hardening gear teeth. The induction coil can be tailored to suit the profile of the component. The depth of the case can be controlled by the frequency of the alternating current. The higher the frequency, the nearer to the surface of the component will be the eddy currents resulting in a shallower depth of heating and, therefore, hardening.

Nitriding

This process is used to put a hard, wear-resistant coating on components made from special alloy steels, for example drill bushes. The alloy steels used for this process contain either 1.0 per cent aluminium, or traces of molybdenum, chromium and vanadium. Nitrogen gas is absorbed into the surface of the metal to form very hard nitrides. The process consists of heating the components in ammonia gas at about 500°C for upwards of 40 hours. At this temperature the ammonia gas breaks down and the atomic nitrogen is readily absorbed into the surface of the steel. The case

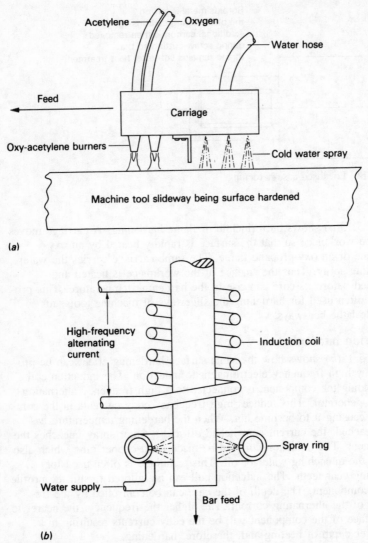

(a)

(b)

Fig. 5.15 Surface-hardening (a) Flame hardening (Shorter process) (b) Induction hardening

is applied to the finished component and no subsequent grinding is possible since the case is only a few micro-metres thick. However, this is no disadvantage since the process does not affect the surface finish of the component and the process temperature is too low to cause distortion. Some of the advantages of nitriding are:

(*a*) cracking and distortion are eliminated since the processing temperature is relatively low and there is no subsequent quenching;

(*b*) surface hardnesses as high as 1150 H_D are obtainable with 'Nitralloy' steels;

(*c*) corrosion resistance of the steel is improved;

(*d*) the treated components retain their hardness up to 500°C compared with the 220°C for case hardened plain carbon and low alloy steels.

6 Cast irons and their heat treatment

6.1 The iron carbon system for cast irons

Cast iron is the name given to those ferrous metals containing more than 1.7 per cent carbon. It is similar in composition to crude pig-iron as produced by the blast furnace. Unlike steel, it is not subjected to an extensive refinement process. After the pig-iron has been remelted in a cupola furnace ready for casting, selected scrap-iron and scrap-steel is added to the melt to give the required composition. Although more expensive, casting ingots can be purchased already formulated to a guaranteed composition and purity where high grade and alloy cast irons are required.

Since pig-iron and scrap are cheap, and because there is no expensive refinement process, cast iron is a low cost material which is useful where castings of high rigidity, resistance to wear, and high compressive strength are required. Further, cast iron is easy to machine, has a high fluidity which makes it easy to cast into intricate shapes, and has a melting point between 1147°C and 1250°C which is substantially lower than the melting point for mild steel.

Reference back to the iron-carbon phase equilibrium diagram (Fig. 4.1) shows that the cast irons lie to the right of the 'steel section' of the diagram. Figure 6.1 shows the 'cast iron section' of the phase equilibrium diagram in greater detail. Since the maximum amount of carbon which can be held in solid solution as austenite is only 1.7 per cent, it is obvious that in all cast irons there will be surplus carbon. This surplus carbon can combine with the iron to form cementite (iron carbide) or it can be precipitated out as flake graphite. (Graphite is an

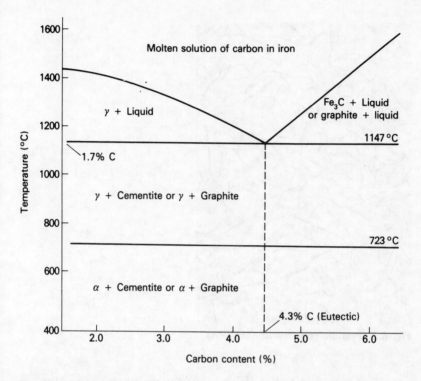

Fig. 6.1 Cast iron section of the iron-carbon phase equilibrium diagram

allotrope of carbon.) In the foundry, cementite is referred to as 'combined carbon' and flake graphite is known as 'free carbon'. Figure 6.1 shows that there is a eutectic when there is 4.3 per cent carbon present. At this composition the molten metal solidifies at 1147°C into austenite (γ phase) and cementite. Unless cooling is very rapid, graphite will be precipitated out due to the instability of the cementite as a result of some of the impurities present (particularly silicon). As cooling proceeds, further graphite is precipitated out from the austenite. At 723°C, providing cooling is slow enough, the remaining austenite (γ phase) changes into ferrite (α phase). Thus at room temperature the composition consists of ferrite plus large flakes of graphite together with fine flakes of graphite formed by the decomposition of the cementite after solidification.

If cooling is sufficiently fast to prevent phase equilibrium from being achieved, the austenite will change to ferrite and pearlite at the eutectoid temperature of 723°C as the cementite essential for the formation of pearlite will be retained. With even faster cooling, the structure will consist of flake graphite in a matrix which is entirely pearlitic as shown in Fig. 6.2. Ferritic and pearlitic cast irons containing free graphite flakes are called *grey cast irons* because of the grey appearance of their surfaces

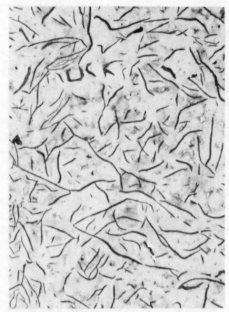

Fig. 6.2 Pearlitic grey cast iron (BCIRA)

when freshly fractured. As in steel, increasing the amount of pearlite present enhances the toughness and hardness of the cast iron.

With even faster cooling still (chilling), a different type of structure is likely to be formed. When the solidus temperature of 1147°C is reached, the structure will consist entirely of austenite and cementite. On further cooling the austenite changes into pearlite. Thus at room temperature the composition of the iron is pearlite and cementite. This type of cast iron is known as *white cast iron* due to its white appearance when freshly fractured. It derives this appearance from the white crystals of cementite. A microphotograph of a typical white cast iron structure is shown in Fig. 6.3. The hardness of the cementite in white cast iron makes it difficult to machine and its use is limited mainly to wear-resistant components and as a basic material for conversion to white-heart malleable cast iron (see Section 6.4).

6.2 Alloying elements and impurities

Cast irons are not solely alloys of iron and carbon as the phase equilibrium diagram would suggest, but complex alloys in which impurities such as sulphur and phosphorus and alloying elements such as silicon and manganese have a significant influence on the properties of

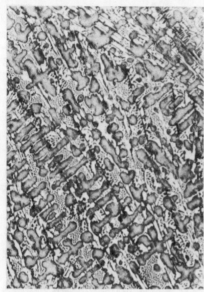

Fig. 6.3 White cast iron (BCIRA)

the casting. Although complex alloys, such cast irons are still referred to as common cast irons, the term *alloy cast irons* being reserved for those cast irons containing substantial amounts of such metallic elements as chromium, nickel, etc.

Silicon

This element is used to *soften* cast irons by promoting the formation of flake graphite at the expense of cementite. The silicon content is increased in irons used for light or thin components which might chill harden by cooling too quickly in the mould, becoming hard and brittle. The addition of significant amounts of silicon can reduce the eutectic composition down to 3.5 per cent carbon. Thus, at a constant rate of cooling, the addition of silicon to a cast iron having 3 per cent carbon will have the following effects.

(*a*) Ferritic grey cast iron is produced with 3 per cent silicon.
(*b*) Ferritic/pearlitic cast iron is produced with 2 per cent silicon.
(*c*) Pearlitic cast iron is produced with 1.5 per cent silicon.
(*d*) White cast iron is produced with no silicon.

Providing sufficient silicon has been added to break down the cementite into flake graphite (3 per cent silicon in this example) there is no benefit in adding extra silicon. In fact, the presence of excess silicon will lead to increased brittleness.

Sulphur

Sulphur is an impurity carried over from the coke used in the blast furnace during the extraction of the iron from its ore. The presence of sulphur in cast irons, even in small quantities, has the effect of stabilising the cementite and preventing the formation of flake graphite. Thus sulphur *hardens* the cast iron. It also causes embrittlement due to the formation of iron sulphide (FeS) at the grain boundaries.

Manganese

The addition of this element in small quantities is essential in all ferrous metals as it combines with any residual sulphur present to form manganese sulphide (MnS). Unlike ferrous sulphide, manganese sulphide is insoluble in the molten iron and floats to the top of the melt to join the slag. Thus by removing the sulphur, the manganese indirectly softens the cast iron and also removes a source of embrittlement. Increasing the manganese content beyond that required to neutralise the sulphur has the effect of stabilising the cementite and causing hardness in the iron, just as the sulphur did, but without any embrittlement. Thus it is important to balance the amount of manganese, sulphur and silicon present with great care. Manganese also promotes grain refinement and increases the strength of the cast iron.

Phosphorus

Like sulphur, this is a residual impurity from the extraction process. It is present in cast iron as iron phosphide (Fe_3P). This phosphide forms a eutectic with ferrite in grey cast irons, and with ferrite and cementite in white cast irons. Since these eutectics melt at only 950°C, high phosphorus content cast irons have great fluidity. Cast irons containing one per cent phosphorus are thus very suitable for the production of thin section castings and highly intricate ornamental castings. Unfortunately phosphorus, like sulphur, causes hardness and embrittlement in the cast iron, and the amount present must be kept to a minimum in castings where shock resistance and strength is important. The composition and applications of some typical grey cast irons are summarised in Table 6.1.

6.3 Heat treatment of grey cast iron

It is virtually impossible to arrange for all the parts of a casting to cool and solidify at the same rate. It is equally impossible to achieve the optimum cooling rate for a given composition of cast iron in any given casting. Therefore grey iron castings are frequently subjected to heat treatment to relieve stresses, refine the grain structure and improve its machinability.

Annealing

Grey iron castings can be heated to just above the Ac_1 temperature (approximately 760°C), soaked at that temperature until equilibrium struc-

Table 6.1 Typical grey cast irons

Composition (%)					Applications
C	Si	Mn	S	P	
3.30	1.90	0.65	0.08	0.15	Motor vehicle brake drums
3.25	2.25	0.65	0.10	0.15	Motor vehicle cylinder blocks
3.25	1.75	0.50	0.10	0.35	Medium machine castings
3.25	1.25	0.50	0.10	0.35	Heavy machine castings
3.60	1.75	0.50	0.10	0.80	Light and medium spun cast water pipes
3.50	2.75	0.50	0.10	0.9	Ornamental castings requiring maximum fluidity but only low strength

tures have been achieved and then cooled very slowly. The soaking time will depend upon the mass and thickness of the casting. This promotes the precipitation of flake graphite and breaks down any excess cementite which may be forming 'hard spots' or *chills*. These hard spots result from over-rapid cooling of thin sections due to the chilling effect of the mould. Annealing also removes internal stresses in the composition resulting from uneven cooling in the casting process.

Quench hardening

Ferritic grey cast irons can be heated to just above the Ac_1 temperature (approximately 760°C) and quenched in water or oil. This prevents phase equilibrium being attained and results in the formation of pearlite rather than ferrite and flake graphite, thus giving increased toughness and hardness. Great care has to be taken in the quenching process to avoid a white iron structure. To relieve the stresses created by quenching it is usual to temper the castings at 450°C to 475°C. Quench hardening grey iron castings is not a common practice.

Stress relieving

Most iron castings have internal stresses when they are released from the mould. These stresses not only reduce the strength of the casting, they may also cause it to warp and distort during machining. Traditionally machine beds and frames which were required to be dimensionally and geometrically stable were 'weathered', that is, they were rough machined and then left out of doors for a long period of time varying from months to years (see also Section 5.6). The continual changes in temperature caused diffusion of the internal structure coupled with continual expansion and contraction. This resulted in the release of the internal stresses and allowed the castings to warp to their final shape so that after finish machining no further distortion took place.

Nowadays, castings are more likely to be stress relieved by soaking them in a furnace at approximately 550°C for a period ranging from several hours to several days depending upon the size of the casting. Very slow cooling follows the heating. Thus the temperature for the stress relief of iron castings is very much lower than that required for the full annealing of steel castings and forgings (see Section 5.5).

6.4 Malleable cast iron

Malleable cast irons are produced from white cast irons by a variety of heat-treatment processes depending upon the final composition and structure required. As their name implies, malleable cast irons have increased malleability and ductility, increased tensile strength, and increased toughness.

Black-heart process

In this process the white iron castings are heated in airtight boxes out of contact with air at 850°C to 950°C for 50 to 170 hours depending upon the mass and thickness of the castings. The effect of this prolonged heating is to break down the iron carbide (cementite) of the white cast iron into small rosettes of graphite. The final structure is of ferrite and fine carbon particles as shown in Fig. 6.4. The name 'black-heart' comes from the darkened appearance of the iron when fractured, resulting from the formation of free graphite. The relationship of the process to the iron-carbide phase equilibrium diagram is shown in Fig. 6.5.

White-heart process

In this process the castings are packed into airtight boxes with iron oxide in the form of high grade ore. They are then heated to about 1000°C for between 70 and 100 hours depending upon the mass and thickness of the castings. The ore oxidises the carbon in the castings and draws it out, leaving a ferritic structure near the surface and a pearlitic structure near the centre of the casting. There will also be some fine rosettes of graphite. White-heart castings behave much as expected of a mild steel casting, but with the advantage of a very much lower melting point and higher fluidity at the time of casting. Figure 6.6 shows a typical white-heart structure.

Pearlitic process

This process is similar to the black-heart process inasmuch as the castings are heated to 850°C to 950°C for 50 to 170 hours in a non-oxidising

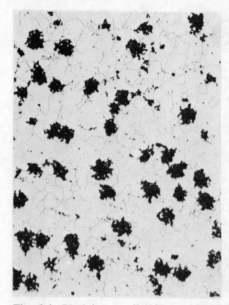

Fig. 6.4 Black-heart malleable cast iron (BCIRA)

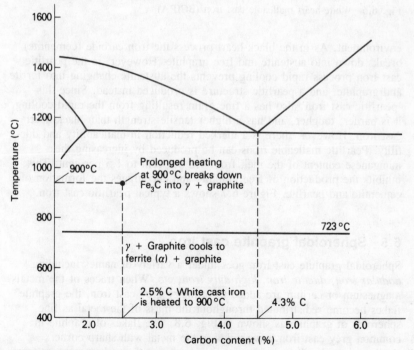

Fig. 6.5 Black-heart transformations

132

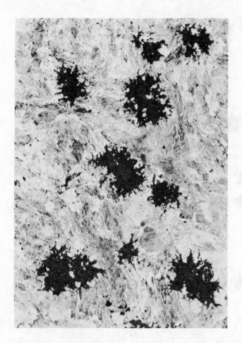

Fig. 6.6 White-heart malleable cast iron (BCIRA)

environment. As in the black-heart process the iron carbide (cementite) breaks down into austenite and free graphite. However, in the pearlitic cast iron process rapid cooling prevents the austenite changing into ferrite and graphite, and a pearlitic structure is produced instead. Since this 'pearlitic cast iron' also has a fine grain resulting from the rapid cooling, it is harder, tougher, and has a higher tensile strength than black-heart cast iron. However, there is a marked reduction in malleability and ductility. Pearlitic malleable irons can be produced by increasing the manganese content of the melt from 1.0 per cent to 1.5 per cent. This inhibits the production of free graphite and encourages the formation of cementite and pearlite. Figure 6.7 shows a typical pearlitic cast iron.

6.5 Spheroidal graphite cast iron

Spheroidal graphite cast iron goes under a variety of names including *nodular iron, ductile iron, high duty iron*, etc. When traces of the metals magnesium or cerium are added to ordinary grey cast iron, the graphite flakes become redistributed throughout the mass of the metal as fine spheroids of graphite as shown in Fig. 6.8. The flakes of graphite in common grey cast iron leave voids in the metal with sharp corners similar to cracks. Stress concentrations occur at these sharp corners from

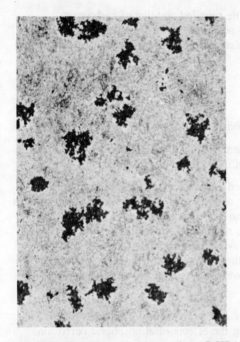

Fig. 6.7 Pearlitic malleable cast iron (BCIRA)

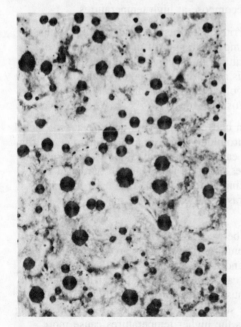

Fig. 6.8 Spheroidal graphite cast iron (BCIRA)

which cracks are propagated. The rounded nodules of carbon in spheroidal graphite cast iron do not create stress concentrations to the same extent, and this greatly enhances the strength of the castings. It also reduces the likelihood of fatigue failure.

6.6 Alloy cast irons

The alloying elements in cast irons are similar to those in alloy steels.

Nickel is used for grain refinement, to add strength, and to promote the formation of free graphite. Thus it toughens the casting.

Chromium stabilises the combined carbon (cementite) present and thus increases the hardness and wear resistance of the casting. It also improves the corrosion resistance of the casting, particularly at elevated temperatures. As in alloy steels, nickel and chromium tend to be used together. This is because they have certain disadvantages when used separately which tend to offset their advantages. However when used together the disadvantages are overcome, whilst the advantages are retained.

Copper is used very sparingly as it is only slightly soluble in iron. However, it is useful in reducing the effects of atmospheric corrosion.

Vanadium is used in heat-resisting castings as it stabilises the carbides and reduces their tendency to decompose at high temperatures.

Molybdenum dissolves in the ferrite and, when used in small amounts (0.5 per cent), it improves the impact strength of the casting. It also prevents 'decay' at high temperatures in castings containing nickel and chromium. When molybdenum is added in larger amounts it forms double carbides, increases the hardness of castings with thick sections, and also promotes uniformity of the microstructure.

Martensitic cast irons contain between 4 per cent and 6 per cent nickel and approximately 1 per cent chromium. For example 'Ni-hard' cast iron. It is naturally martensitic in the cast state but, unlike alloys with rather less nickel and chromium, it does not need to be quench hardened thus reducing the possibility of cracking and distortion. It is used for components which need to resist abrasion. It can only be machined by grinding.

Austenitic cast irons contain between 11 per cent and 20 per cent nickel and up to 5 per cent chromium. These alloys are corrosion-resistant, heat-resistant, tough, and non-magnetic. Since the melting temperatures of alloy cast irons can be substantially higher than those for common grey cast irons, care must be taken in the selection of moulding sands and the preparation of the surfaces of the moulds. Increased venting of the moulds is also required as the higher temperatures cause more rapid generation of steam and gases. The furnace and crucible linings must also be suitable for the higher temperatures and the inevitable increase in maintenance costs is also a significant factor when working with high alloy cast irons.

The *growth* of cast irons is caused by the breakdown of pearlitic cementite into ferrite and graphite at approximately 700°C. This causes an increase in volume. This increase in volume is further aggravated by hot gases penetrating the graphite cavities and oxidising the ferrite grains. This volumetric growth causes warping and the setting up of internal stresses leading to cracking, particularly at the surface. Therefore, where castings are called upon to operate at elevated temperatures, alloy cast irons should be used. A low cost alloy is 'Silal' which contains 5 per cent silicon and a relatively low carbon content. The low carbon content results in a structure which is composed entirely of ferrite and graphite with no cementite present. Unfortunately 'Silal' is rather brittle because of the high silicon content. A more expensive alloy is 'Nicrosilal'. This is an austenitic nickel-chromium alloy which is much superior in all respects for use at elevated temperatures.

6.7 Properties and uses of white cast irons

This is little used except as a basis for malleable cast irons (Section 6.4). It is wear resistant but lacks ductility and breaks easily since it is very brittle. It is virtually unmachineable except by grinding. The composition and properties of a typical white cast iron may be listed as follows:

Carbon	2.5%	Elongation:	nil%
Silicon	0.8%	Tensile strength:	250 to 450 MPa
Manganese	0.4%	Hardness:	400 H_B
Sulphur	0.08%		
Phosphorus	0.1%		

6.8 Properties and uses of grey cast irons

These are the most widely used cast irons and they vary in composition according to specific applications. Table 6.1 has already listed some typical examples together with their composition and uses. The tensile strength of such cast irons varies between 150 MPa and 350 MPa.

Cast irons requiring high fluidity are used for very intricate mouldings, for decorative iron work and architectural tracery. They lack mechanical strength since the high fluidity is obtained by ensuring that the melt has a high silicon content (2.5% to 3.5%) and a high phosphorus content (approximately 1.5%). It is the high phosphorus content which reduces the strength of the metal.

General purpose engineering irons have to have reasonable mechanical strength. Ideally they should have fine graphite flakes in a matrix of pearlite. Such a cast iron would combine good mechanical properties with good machineability. The silicon content would be dependent upon the thickness and mass of the casting; varying between 2.5 per cent for thin sections to about 1.5 per cent for castings with thick sections. The phosphorus content would be kept low to improve toughness and shock resistance, although up to 0.8 per cent may be present to improve fluidity. Sulphur must be kept below 0.1 per cent to avoid segregation, hard spots and embrittlement.

Local hardening of grey iron castings (e.g. slideways) can be achieved by *chilling*. This is done by introducing 'chills' or metal plates into the mould just behind the surface layer of sand to promote rapid cooling and the formation of cementite. The hardness occurs only at the surface of the casting and the core remains grey and tough. (See also *flame-hardening* (Section 5.14).)

Large heavy castings do not require a high silicon content as they naturally cool slowly and there is adequate time for any cementite to break down into ferrite and flake graphite. A typical composition would be 1.2 per cent to 1.5 per cent silicon, 0.5 per cent phosphorus, and 0.1 per cent sulphur. Table 6.2 (derived from BS 1452) lists the essential properties of some typical grey cast irons.

6.9 Specifications for grey iron castings

BS 1452 specifies the requirements for seven grades of grey iron castings. These are: 150, 180, 220, 260, 300, 350 and 400. As well as designating the specification, these numbers refer to the tensile strength of a standard test-piece in MPa (remember that 1 MPa = 1 MN/m^2 = 1 N/mm^2) taken from the same melt as the castings being manufactured. The standard does not specify the chemical composition of the iron nor its processing in the foundry. It only specifies the properties, test conditions and quality. How these are attained is left to the discretion of the foundry in consultation with the customer. Basically, the customer need only specify the standard and the grade, e.g. BS 1452 grade 220, in which case the foundry will then use its discretion in achieving this specification. However the customer may also specify or require:

(*a*) a mutually agreed chemical composition;
(*b*) casting tolerances, machining locations;
(*c*) test bars and/or test certificates;
(*d*) whether testing and inspection is to be carried out in the presence of the customer's representative;
(*e*) any other requirement such as hardness tests and their locations, nondestructive tests, and quality assurance.

Throughout BS 1452 the accent is on the quality of the finished casting and the methods of testing to ensure that the specified quality and properties have been achieved. For example, the frequency of testing is specified as follows.

Table 6.2 Properties of grey cast iron

Grade	UTS (MPa)	0.1% Proof stress (MPa)	Compressive Strength (MPa)	0.1% Compressive Proof Stress (MPa)	Shear Strength (MPa)	Modulus of Elasticity (GPa) Tension	Compn.	Modulus of Rigidity (GPa)
150	150	98	600	195	173	100	100	40
180	180	117	672	234	207	109	109	44
220	220	143	768	286	253	120	120	48
260	260	169	864	338	299	128	128	51
300	300	195	960	390	345	135	135	54
350	350	228	1080	455	403	140	140	56
400	400	260	1200	520	460	145	145	58

For grey cast iron, hardness is not related to tensile strength but varies with casting section thickness and materials composition.

Grades 150 and 180: one test bar for up to 10 tonnes of castings.
Grades 220 and 260: one test bar for up to 5 tonnes of castings.
Grades 300, 350 and 400: one test bar for up to 1 tonne of castings.

Where a single casting exceeds the above figures, there shall be one test bar per casting. The test bar should be sufficient to manufacture at least three test-pieces to allow for retesting.

BS 1452 also specifies that tensile testing (Section 11.2) shall be in accordance with BS 18: Part 2 and that the test-piece shown in Fig. 6.9 shall be machined from a cast cylindrical test bar of 30 mm diameter and 230 mm minimum length. At least three test-pieces should be cast so that in the event of the first test-piece failing, two further tests can be made. Should either of these retests fail, then the whole melt and any castings made from it is deemed to have failed the requirements of BS 1452.

Although the BS grades of grey cast iron are based upon minimum tensile strengths obtained from test-pieces machined from 30 mm diameter cast test bars, the strength of the castings produced will vary according to their mass, section thickness and consequential rate of cooling. Thin sections of low mass will cool quickly and show increased tensile strength and hardness, whilst thicker sections will cool more slowly and show a reduced tensile strength and hardness. This effect is shown in Fig. 6.10.

Many other properties of the grey cast irons specified in BS 1452 have to be considered and the full specifications are comprehensively listed in Appendix B of that specification.

6.10 Properties and uses of malleable cast irons

Malleable irons exploit the excellent casting properties of cast iron during the casting process, after which they are converted by heat treatment processes into a composition and structure whose properties resemble that of a low-carbon steel. This results in castings which are stronger and which are much less brittle than ordinary cast irons, and are widely used in the automobile, agricultural machinery, and machine tool industries, for the manufacture of small and medium sized stressed components. Malleable iron castings are also used in the electrical industry for conduit fittings, switch gear cases and components. The properties and applications of some typical malleable cast irons (derived from BS 6681) are listed in Table 6.3.

6.11 Specifications for malleable cast irons

BS 6681 supersedes BS 309, 310 and 3333 for specifying the requirements of malleable cast irons. These are specified by a capital letter designating the type of iron:

W = whiteheart malleable cast iron;
B = blackheart malleable cast iron;
P = pearlitic malleable cast iron.

Gauge diameter d		Minimum parallel length L_c	Minimum radius r		Plain ends		Screwed ends	
Nominal value	Machining tolerance*		Nominal value	Machining tolerance	Minimum diameter D_1	Minimum length L_p	Minimum diameter at root of thread D_2	Minimum length L_s
mm	mm	mm	mm	mm	mm	mm	mm	mm
20	± 0.5	55	25	−0, +5	25	65	25	30

*If it is desired to calculate the tensile strength on the basis of the nominal diameter, the machining tolerance shall be ± 0.10 mm.

NOTE. With screwed ends, any form of thread may be used provided that the diameter at the root of the thread is not less than that specified

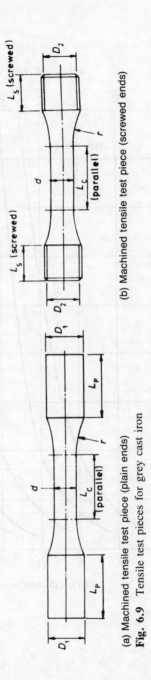

(a) Machined tensile test piece (plain ends)

(b) Machined tensile test piece (screwed ends)

Fig. 6.9 Tensile test pieces for grey cast iron

140

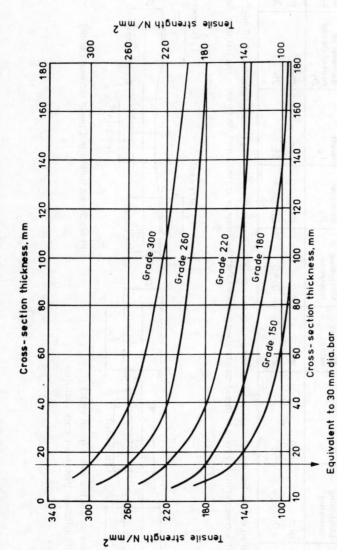

Example. An iron which will give a tensile strength of 220 N/mm² in a 30 mm diameter test bar should give a tensile strength of approximately 220 N/mm² in the centre of a casting whose ruling section is 15 mm in thickness and approximately 147 N/mm³ where the cross section is 100 mm in thickness.

Variation of tensile strength with section thickness.

Fig. 6.10 Variation of tensile strength with section thickness

Courtesy B.S.I. BS1452: 1977

Table 6.3 Properties and uses of malleable cast iron

Type of cast iron	Condition	Properties			Applications
		U.T.S. (MPa)	Elong^n (%)	Hardness (H_B)	
Blackheart malleable	Annealed	300−350	6−12	150 max.	Wheel hubs, brake drums, conduit fittings, control levers and pedals.
Whiteheart malleable	Annealed	340−480	3−15	230 max.	Wheel hubs, bicycle and motor cycle frame fittings. Gas, water, and steam pipe fittings.
Pearlitic mealleable	Normalised	450−700	3−6	150−290	Gears, couplings, camshafts, axle housings, differential housings and components.

This letter is followed by two figures designating the minimum tensile strength in MPa of a 12 mm diameter test-piece. For this purpose the tensile strength is divided by ten, e.g. if the tensile test returned a result of 320 MPa, the two figure designation would be 32. Finally, there are two figures representing the minimum elongation percentage on the specified gauge length (see Fig. 6.11). Thus a complete designation for a malleable cast iron could be:

W 35 − 04 = Whiteheart malleable cast iron with a minimum tensile strength of 350 MPa, and a minimum elongation of 4%.

As for grey cast irons the specification is not concerned with the chemical composition except for stating that the phosphorus content shall not exceed 0.12%. The composition and manufacturing processes are left to the discretion of the foundry in consultation with the customer. The melt and the castings made from it will satisfy the requirements of BS 6681 providing the test results and the general quality of the castings meets the specifications laid down.

There should be at least one test bar produced from each melt and at least one test bar for each 2000 kg of castings produced. Two additional

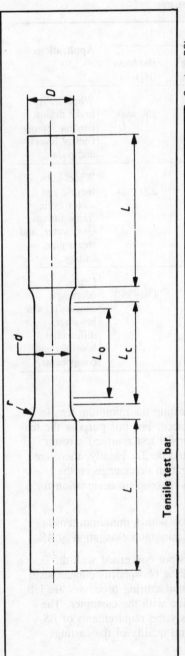

Tensile test bar

Courtesy BSI
BS6681: 1986

Dimensions of tensile test bar

Diameter (d) mm	Tolerance on diameter mm	Nominal cross-sectional area, (S_0) mm²	Gauge length ($L_0 = 3d$) mm	Minimum parallel length (L_c) mm	Radius at shoulder (r) mm	Preferred shank dimensions (for information)	
						Diameter (D) mm	Length (L) mm
9	± 0.7	83.6	27	30	6	13	40
12	± 0.7	113.1	36	40	8	16	50
15	± 0.7	176.7	45	50	8	19	60

NOTE. Where the test bar is tested in the unmachined state, the tensile strength is calculated from the measured diameters of each test bar. For this purpose, the diameter is obtained by taking the average of measurements taken in the same plane at right angles to each other.

Fig. 6.11 Tensile test pieces for malleable cast iron

test bars from each batch should be kept in reserve for retests (see Section 6.9). Since malleableising is a heat treatment process, the test bars should be heat treated alongside the castings to which they refer.

BS 6681 sets out the full range of mechanical properties for each designated grade of malleable cast iron and a comprehensive testing programme to ensure that the specifications are met. These details are beyond the scope of this chapter and the specification itself should be consulted.

6.12   Properties and uses of spheroidal and nodular graphite cast irons

Spheroidal graphite cast irons are now used for quite highly stressed components in the automobile industry. By designing components with this material in mind, spheroidal graphite castings can, in some instances, replace steel forgings at much lower cost, for example in camshafts, crankshafts, and differential gear carriers for motor vehicles. The properties and some further applications for typical spheroidal graphite (SG) cast irons are listed in Table 6.4.

6.13   Specifications for spheroidal and nodular graphite cast irons

BS 2789 specifies the requirements for spheroidal or nodular graphite cast irons. Again, the standard does not specify the chemical composition of the iron, its method of manufacture or any subsequent heat treatment. The standard is solely concerned with the properties, testing and quality control of the finished castings. How this is attained is left to the discretion of the foundry in consultation with the customer. It is a very comprehensive standard and it is only possible to review briefly some of its more important points within the scope of this chapter. The standard itself should be consulted for more detailed study.

This revised standard includes requirements for tensile strength, elongation, 0.2% proof stress and, for two grades of iron, resistance to impact. The standard covers the majority of commercial applications and includes the requirements for a total of nine grades of iron including two new grades: 900/2 which increases the range of mechanical properties available to designers, and grade 450/10 which has a higher proof stress to tensile stress ratio than the previous grades. The grades specified are:

900/2   This grade has a tempered martensitic structure.

800/2   These grades have a mainly pearlitic matrix, characterised by
700/2   high tensile strength but at the expense of lower ductility and
600/3   resistance to impact.

500/7   These intermediate grades have ferritic/pearlitic matrices

450/10   combining strength with reasonable ductility and impact
           strength.

420/12   This grade has a mainly ferritic matrix of moderately high
           tensile strength, but with subsequent ductility and impact
           resistance.

400/18   These grades have wholly ferritic matrices with even greater
350/20   ductility and even higher resistance to impact.

To interpret these grades, the first three figures indicate the minimum
tensile strength in MPa and the final figure (after the /) indicates the
minimum elongation percentage. The addition of the letter L followed by
a number in the case of 400/18L20 and 350/22L40 indicates that the
impact strength must be attained at low temperatures, that is at $-20°C$
and $-40°C$ respectively.

The standard specifies the shape and dimensions of test-pieces for tensile testing and it also specifies how the test bars are derived, that is
whether they are cast separately or whether they are cast on to the main
casting or on to a runner bar. If they are 'cast on', then they must not be
separated from the main casting until they have cooled below 500°C.

6.14   Composition, properties and applications of some alloy cast irons

Although much more expensive because of the high cost of the alloying
elements which they contain and the increased foundry costs in casting
them, they are widely used for special applications where their properties
can be exploited. By suitable alloying, their properties can be tailored to
give high strength, high corrosion resistance, high wear resistance, and
high temperature resistance, whilst retaining the improved casting properties of cast iron compared with cast steel. They can also be cast to
intricate shapes which cannot be achieved by steel forgings whilst
exhibiting almost comparable mechanical properties. The composition,
properties and some typical applications of alloy cast irons are listed in
Table 6.5.

Table 6.4 Properties and uses of spheroidal graphite cast iron

Type of cast iron	Grade	Properties				Applications
		UTS (MPa)	0.2% Proof Stress	Elong^n %	Hardness (H_B)	
Ferritic	350/22	350–420	220–270	12–22	160–212	Water main pipes, hydraulic cylinder and valve bodies, machine vice handles.
Ferritic/Pearlitic	450/10 to 600/3	450–600	320–370	3–10	160–269	
Pearlitic	700/2	700–800	420–480	2	229–352	These grades will surface harden and can replace steel forgings for such stressed applications as automobile engine camshafts and crankshafts.
Martensitic	900/2	900	600	2	302–359	

Table 6.5 Typical alloy cast irons

Type of cast iron	Composition (%)								Properties		Applications
	C	Si	Mn	S	P	Ni	Cr	Other Elements	U.T.S. (MPa)	Hardness (H$_B$)	
Chromidium	3.2	2.1	0.8	0.05	0.17	—	0.32	—	275	230	Cylinder blocks, brake drums and discs, clutch casings, differential carriers, etc.
Wear & shock resistant	2.9	2.1	0.7	0.05	0.10	1.75	0.10	0.8 Mo 0.15 Cu	450	300	Crankshafts for automobile diesel and petrol engines. High strength. Good shock and vibration damping properties.
Ni-hard	2.8	1.3	—	—	—	21.0	2.0	—	—	60	Martensitic iron of great hardness and wear resistance, ore crushing jaws, abrasive material handling components.
Ni-Resist	2.9	2.1	1.0	0.05	0.1	15.0	2.0	6.00 Cu	215	130	A corrosion resistant alloy suitable for valve and pump bodies handling sulphur and chloride solutions.
Silal	2.5	5.0	—	—	—	—	—	—	215	—	High temperature resistance, suitable for exhaust manifolds, furnace components, etc.

7 Non-ferrous metals, their alloys, and their heat treatment

7.1 Non-ferrous metals

The term 'non-ferrous metals' refers to the thirty-eight metals other than iron that are known to man. The non-ferrous metals which are most commonly used by engineers are listed in Table 7.1. Two of the most important non-ferrous metals are aluminium and copper. They not only form the bases of many important alloys, but they are widely used in their own right as pure metals.

A list of non-ferrous metals would not be complete without mention of the 'new metals' listed below. Although known for many years, these metals have only been available in bulk for engineering applications since the Second World War. Further, it is only comparatively recently, with the development of supersonic aircraft and the nuclear power industry, that there has been a large scale commercial demand for these materials.

Niobium, tantalum, zirconium: atomic reactor components.
Tellurium: used instead of lead in free-cutting alloys (higher strength).
Titanium: used in supersonic aircraft and rockets as it has a higher strength/weight ratio than aluminium alloys and retains its strength at the elevated temperatures met with in supersonic flight.
Beryllium: used as an alloying element with copper to make instrument springs which are resistant to fatigue and corrosion. Beryllium-copper alloys can be quench hardened like steel and are used to provide 'non-sparking' tools for use in oil fields, chemical installations, gas rigs, etc.

Table 7.1 Common non-ferrous metals

Metal	Density (kg/m^3)	Melting point (°C)	Properties	Typical uses
Aluminium	2700	660	Lightest of the commonly used metals. High electrical and thermal conductivity. Soft, ductile and low tensile strength 93 MPa	The base of many engineering alloys. Lightweight electrical conductors
Copper	8900	1083	Soft, ductile and low tensile strength 232 MPa. Second only to silver in conductivity, it is much easier to joint by soldering and brazing than aluminium. Corrosion resistant	The base of brass and bronze alloys. It is used extensively for electrical conductors and heat exchangers, such as motor car radiators
Lead	11 300	328	Soft, ductile and very low tensile strength. High corrosion resistance	Electric cable sheaths. The base of 'solder' alloys. The grids for 'accumulator' plates. Lining chemical plant. Added to other metals to make them 'free-cutting'
Silver	10 500	960	Soft, ductile and very low tensile strength. Highest electrical conductivity of any metal	Widely used in electrical and electronic engineering for switch and relay contacts
Tin	7300	232	Resists corrosion	Coats sheet mild steel to give 'tin plate'. Used in soft solders. One of the bases of 'white metal' bearings. An alloying element in bronzes
Zinc	7100	420	Soft ductile and low tensile strength. Corrosion resistant	Used extensively to coat sheet steel to give 'galvanized iron'. The base of

Metal	Density (kg/m³)	Melting point (°C)	Properties	Typical uses
				die-casting alloys. An alloying element in brass
Chromium	7500	1890	Resists corrosion. Raises strength but lowers ductility of steels. Improves heat-treatment properties	Used as an alloying element in high-strength and corrosion-resistant steels. Used for electro-plating
Cobalt	8900	1495	Improves wear-resistance and 'hot hardness' of high-speed steels	Used as an alloying element in 'super' high-speed steels and in permanent-magnet alloys
Manganese	7200	1260	High affinity for oxygen and sulphur. Soft and ductile	Used to de-oxidise steels and to offset the ill-effects of the impurity sulphur. Larger amounts improve wear resistance
Molybdenum	9550	2620	A heavy, heat-resistant metal that alloys readily with other metals	Used as an alloying element in high-strength nickel-chrome steels to improve mechanical and heat-treatment properties. It reduces mass effect and temper-brittleness
Nickel	8900	1458	A strong, tough, corrosion-resistant metal widely used as an alloying element	Used as an alloying element to improve the strength and mechanical properties of steel. Tends to unstabilise the carbon during heat-treatment, and chromium has to be added to counteract this effect in medium and high-carbon steels. Used for electro-plating

These 'new metals' are very expensive compared with the more conventional engineering materials and they are only used where their special properties can be fully exploited.

The pure non-ferrous metals are used mainly where their properties of corrosion resistance and high electrical and thermal conductivity can be exploited. They are not widely used as structural materials in mechanical engineering because of their relatively low strengths. However, as will be shown later in this chapter, their mechanical properties can be greatly enhanced by alloying these metals together.

7.2 Aluminium

Pure aluminium is a weak, ductile material with a low density (2.3 g mm^{-3} compared with 7.9 g mm^{-3} for iron). Its electrical conductivity is second only to copper as is its thermal conductivity. It is also highly resistant to atmospheric corrosion. Because of its low strength it is of little use as a structural material, and for such purposes aluminium alloys are preferred.

High purity aluminium

This is used where corrosion resistance or high electrical or high thermal conductivity is required. The impurities present are less than 0.5 per cent. Aluminium has a high affinity for atmospheric oxygen and a film of aluminium oxide quickly forms on a freshly cut surface. This film is virtually homogeneous and prevents further corrosion taking place. The oxide film is also virtually transparent, so the aluminium retains its surface appearance for a long time. Unfortunately aluminium reacts violently with alkalis to give off hydrogen, so that care has to be taken not to subject it to caustic degreasing compounds. Neither is high purity aluminium suitable for marine environments — special alloys have been developed for marine use. Aluminium is used for lining food processing vessels, and also for architectural purposes both internally and externally.

Aluminium is a good conductor of electricity. It is second only to copper with a conductivity of approximately two thirds of that metal. However, because of its lower density, it is a better conductor on a weight for weight basis. For this reason it is used for the conductors in the overhead grid system, where aluminium conducting strands are layed up over a high tensile steel core to form a composite cable. Aluminium also has a high thermal conductivity, but its use in lightweight heat exchangers is limited because of the difficulties encountered in soldering, brazing or welding it on a production basis.

Commercial purity aluminium

Commercially pure aluminium contains from 1.0 to 0.5 per cent impurities and these have the effect of strengthening the metal at the expense of reducing its corrosion resistance and electrical conductivity.

Table 7.2 Properties of Aluminium

Type	Condition	U.T.S. (MPa)	Elong^n (%)	Hardness (H_B)
High purity (99.99% Al)	Annealed	45	60	15
	Half-hard	82	24	22
	Full-hard	105	12	30
Commercial purity (99.0% Al)	Annealed	87	43	22
	Half-hard	120	12	35
	Full-hard	150	10	42

NOTE: High-purity aluminium has superior electrical conductivity and corrosion-resistance properties.

Table 7.2 compares the properties of high purity and commercial purity aluminium. It can be seen from the table that the mechanical properties are heavily influenced by the amount of *cold-working* that the metal has received. By controlling the amount of cold-working, varying degrees of strength and hardness can be produced. These are said to be the different *tempers* in which the metal is available. Since only a few of the non-ferrous alloys — and none of the non-ferrous metals — can be hardened by heat-treatment processes, the modification of the mechanical properties by cold-working becomes an important consideration. The properties, uses, and forms of supply of aluminium are summarised in Table 7.3.

7.3 Aluminium alloys

Aluminium alloys can be divided up into four categories as shown below:

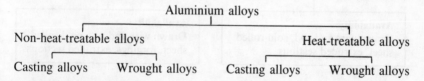

Non-heat-treatable alloys, as their name implies, do not respond significantly to heat-treatment processes beyond annealing and stress relief after casting or cold-working.

Heat-treatable alloys, as their name implies, do respond to heat treatment and in particular to the processes known as solution treatment and precipitation hardening (see Section 7.5).

Casting alloys are alloys which can be used successfully for casting by a variety of processes including sand-casting and die-casting. They may be heat-treatable or non-heat-treatable.

Table 7.3 Aluminium

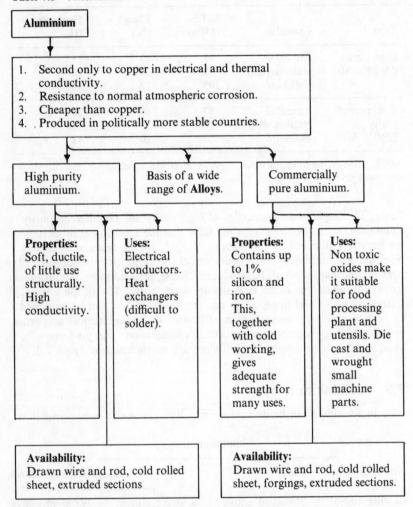

Wrought alloys are those alloys whose mechanical properties allow them to be formed by a variety of processes including forging, rolling, extrusion and drawing. They may be heat-treatable or non-heat-treatable. The non-heat-treatable wrought alloys can have their mechanical properties enhanced considerably by a combination of hot-working and cold-working.

7.4 Aluminium alloys (non-heat-treatable)

Casting alloys

The non-heat-treatable casting alloys are essentially binary alloys containing aluminium and silicon. It can be seen from the phase equilibrium

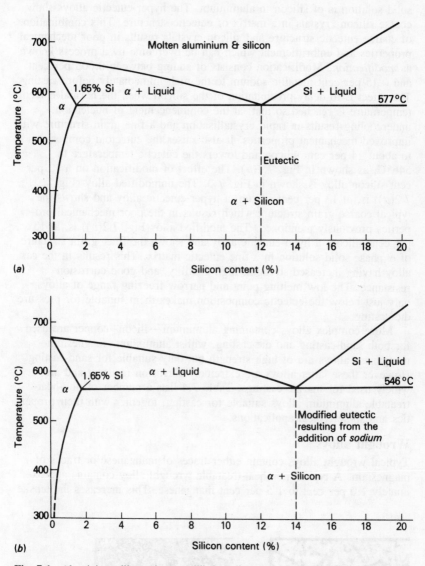

Fig. 7.1 Aluminium-silicon phase equilibrium diagrams (*a*) 'Unmodified' aluminium-silicon phase equilibrium diagram (*b*) 'Modified' aluminium-silicon phase equilibrium diagram

diagram shown in Fig. 7.1(*a*) that silicon is only partially soluble in aluminium below the eutectic composition of approximately 12 per cent silicon. Thus the diagram is of the combination type. The addition of silicon to aluminium increases its fluidity and improves its general casting properties.

The microstructure for the eutectic composition shows a coarse eutectic structure of the α phase solid solution plus silicon. In this alloy the

solid solution is of silicon in aluminium. The hyper-eutectic alloys show
coarse silicon crystals in a matrix of eutectic structure. This combination
of coarse eutectic structure and silicon crystals results in poor mechanical
properties and embrittlement which can be overcome by a process known
as *modification*. Modification consists of adding between 0.005 per cent
and 0.15 per cent metallic sodium to the melt immediately before casting.
The effect is to delay precipitation of the silicon when the normal eutectic
temperature is reached so that, at the commencement of nucleation,
undercooling results in rapid crystallisation and a fine grain structure with
improved mechanical properties. It also raises the eutectoid composition
to about 14 per cent silicon and lowers the eutectic temperature to
546°C, as shown in Fig. 7.1(*b*). The effect of modification on a 13 per
cent silicon alloy is shown in Fig. 7.2. The unmodified alloy (Fig.
7.2(*a*)) is, at 13 per cent silicon, a hyper-eutectic alloy and shows the
typical coarse grain structure which results in the poor mechanical properties previously mentioned. The modified alloy(Fig. 7.2(*b*)) is, at 13
per cent silicon, a hypo-eutectic alloy and shows the finer grain structure
of α phase solid solution in a fine eutectic matrix. This results in the cast
alloy having increased strength, high ductility, and good corrosion
resistance. The low melting point and narrow freezing range of alloys
only just below the eutectic composition makes them suitable for pressure
die-casting.

More complex alloys containing aluminium−silicon−copper are used
for both sand-casting and die-casting, whilst aluminium−magnesium−
manganese alloys are of high strength but only suitable for sand-casting.
However these latter alloys are extremely corrosion resistant and are
widely used for marine castings. Table 7.4 lists a number of non-heat-
treatable aluminium alloys suitable for casting, together with their properties and some typical applications.

Wrought alloys

Typical wrought alloys contain either traces of manganese or traces of
magnesium. A typical non-heat-treatable wrought alloy contains approximately 1.0 per cent to 1.5 per cent manganese. This increases the tensile

Fig. 7.2 Aluminium-silicon cast alloy structures (*a*) 13% silicon cast alloy (unmodified)
(*b*) 13% silicon cast alloy (modified)

Table 7.4 Non-heat-treatable, cast aluminium alloys

Type	Composition		Condition	Properties			Applications
	Cu	Si		0.1% P.S. (MPa)	U.T.S. (MPa)	Elong. (%)	
BS 1490/LM2	1.6	10.0	Chill cast	84	224	2.5	Gravity die-casting. General purpose alloy for lightly stressed parts not subjected to mechanical shock.
BS 1490/LM4	3.0	5.0	Sand cast	70	150	2	Sand-castings; gravity and pressure die-castings. General purpose alloy where mechanical properties are of secondary importance.
			Chill cast	80	170	3	
BS 149/LM6	—	11.5	Sand cast	55	170	7	Sand-castings; gravity and pressure die-castings. Excellent foundry properties. One of the most widely used aluminium alloys when modified. Sumps. Gear boxes, radiators, large castings.
			Pressure die-cast	85	215	4	
Birmalite	(Mn) (0.35)	10.0	Sand cast	98	105	NIL	Sand-castings and gravity die-castings. Maintains its strength. Up to 300 °C, thus used for low-duty pistons and cylinder heads.
			Chill cast	154	168	0.5	

strength without materially affecting the excellent ductility of pure aluminium. This alloy is corrosion-resistant and widely used for kitchen utensils, corrugated sheeting for roof decking panels for the building trade, and for drawn aluminium tubing.

Another important non-heat-treatable wrought-alluminium alloy contains magnesium. This alloy is highly corrosion-resistant, particularly in marine environments, and is widely used in shipbuilding. Unfortunately the low melting point of aluminium-magnesium alloys, compared with steel, produces problems where the alloy is used for fire-resistant bulkheads. Figure 7.3 shows the relationship between tensile strength and magnesium content. It can be seen that the strength increases significantly as the magnesium content is increased to a practical limit of 7 per cent. There is no significant reduction in the ductility of this alloy as its strength increases. Table 7.5 lists examples of non-heat-treatable wrought aluminium alloys, together with their composition, properties and some typical applications.

7.5 Aluminium alloys (heat-treatable)

These are aluminium-copper alloys which, together with other alloying elements, respond to heat treatment. Reference back to Section 3.2 shows that solubility varies with temperature and that when the temperature of a saturated solution is lowered, the excess solute will *precipitate* out of solution. A similar effect can occur in solid solutions and advantage is taken of this in the heat treatment of some non-ferrous alloys, notably those containing aluminium and copper, and those containing aluminium and magnesium.

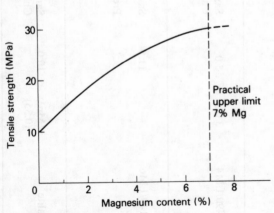

Fig. 7.3 Effect of magnesium content on the tensile strength of annealed aluminium-magnesium alloys

Table 7.5 Non-heat-treatable, wrought aluminium alloys

Type	Composition		Condition	Properties			
	Mg	Other Elements		0.1% P.S. (MPa)	U.T.S. (MPa)	Elong. (%)	Applications
BS 1470/7:N3	(Mn) (1.2)	Cu 0.15 Si 0.6 Fe 0.75	Annealed Hard	45 170	110 200	34 4	Metal boxes, milk bottle caps, food containers, cooking utensils, roofing sheets, panelling of road transport vehicles and railway coaches.
BS 1470/7:N4	2.5	Cu 0.15 Si 0.6 Fe 0.75 Mn 0.5	Annealed $\frac{3}{4}$ hard	75 215	185 265	24 4	Marine superstructures, lifeboats, panelling subjected to marine environments, chemical plant, panelling for road transport vehicles and railway rolling stock.
BS 1470/4:N6	5.0	Cu 0.15 Si 0.6 Fe 0.75 Mn 1.0	Annealed $\frac{1}{4}$ hard	125 215	265 295	18 8	Shipbuilding and applications requiring high strength and corrosion resistance.
BS 1470/4:N7	7.0	Cu 0.15 Si 0.6 Fe 0.75 Mn 1.0	Extruded	175	308	35	Roofing supports in mines and similar applications requiring high strength and corrosion resistance.

158

Solution treatment

Figure 7.4 shows part of the aluminium−copper phase equilibrium
diagram. If an aluminium alloy containing 4 per cent copper is cooled
from the molten condition to 500°C (T_1), the alloy consists of crystals of
α phase solid solution of copper in aluminium. The solid solution does
not become saturated until the solvus is reached at temperature T_2. If
cooling continues slowly (equilibrium conditions) the crystals of solid
solution will remain saturated but some precipitation will occur. At room
temperature (T_3) the structure of the alloy will consist of the α phase
solid solution containing a coarse precipitate of the copper−aluminium
intermetallic compound $CuAl_2$. In this condition the alloy will be soft and
relatively weak.

If, on the other hand, the alloy is quenched from the temperature T_1
so that it cools quickly, there is not time for equilibrium conditions to be
achieved. This is because it requires the grouping together of atoms by
diffusion for precipitation to occur and this is a relatively slow process.
The result of quenching from temperature T_1 is to prevent precipitation so
that the grains consist of a *supersaturated* solid solution of α phase alloy
at room temperature. This is referred to as *solution treatment* and is the
way in which such aluminium alloys may be 'annealed' or softened ready
for cold working. In this condition the alloy will have a fine grain structure and will be somewhat harder, stronger and tougher than when cooled
under equilibrium conditions, but still very ductile.

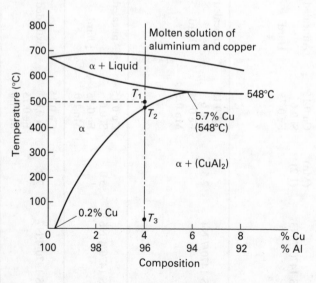

Fig. 7.4 The solution treatment of aluminium-copper alloys

Precipitation hardening

The solution treated alloy can retain its supersaturated solid solution of α phase grains at room temperature. However this is not a stable condition and precipitation of the copper–aluminium intermetallic compound $CuAl_2$ will occur with elapse of time. The precipitate will be in the form of fine particles evenly distributed through the mass of the metal to give greater strength and hardness than that resulting from the coarse precipitation attained by cooling under equilibrium conditions. When precipitation hardening occurs naturally over a period of about 4 days it is referred to as *natural ageing*. Where the precipitation process is speeded up by reheating the alloy to about 165°C for about 10 hours, it is referred to as *artificial ageing*. Both these hardening process are referred to as *precipitation age hardening*. The precipitation process chosen has an appreciable effect on the strength and hardness of the alloy as can be seen from Fig. 7.5. Since solid solutions tend to be soft and ductile (the alloy is softened by 'solution treatment') and intermetallic compounds tend to be hard and brittle, it is the presence of the fine particles of the $CuAl_2$ precipitate which increases the strength and hardness of the precipitation age-hardened alloy. The greater the amount of the in-termetallic precipitate present, the harder and stronger will be the alloy (see also Fig. 5.6).

To retard the precipitation process after solution treatment, components such as rivets made from 'duralumin' alloy are kept under refrigerated conditions in order to avoid the possibility of cracking whilst being cold-worked (e.g. cold-headed).

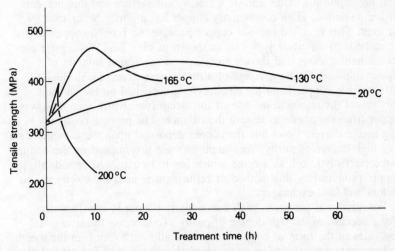

Fig. 7.5 Effects of time and temperature on the precipitation hardening of aluminium alloys

Casting alloys

These are usually complex alloys containing copper or nickel or both in significant amounts plus other alloying elements in lesser amounts. A number of these alloys contain up to 4 per cent copper, whilst others contain up to 2 per cent nickel. They are softened and grain refined after casting by solution treatment and hardened and strengthened by precipitation hardening when intermetallic compounds such as $CuAl_2$ and $NiAl_3$ will be present. These intermetallic compounds make the castings less ductile. Examples of these heat-treatable casting alloys are given in Table 7.6, together with their composition and some typical applications.

Wrought alloys

These are complex alloys of aluminium together with such alloying elements as copper, magnesium, silicon and zinc. One of the most popular of the heat-treatable wrought alloys already mentioned is 'duralumin'. It is strong and tough, yet it can be cold-worked and easily machined after solution treatment. Table 7.7 lists examples of heat-treatable wrought alloys, together with their composition and some typical applications.

7.6 Copper

This is one of the few non-ferrous metals which has sufficient strength to be used unalloyed. Very pure copper has a density of 8.93 g mm^{-3}. It is very ductile and can be readily drawn into rods, wires and tubes. It has excellent electrical and thermal conductivity, and it has good corrosion resistance. Like aluminium it reacts with atmospheric oxygen to form a thin, homogeneous oxide film on a freshly cut surface and this prevents further oxidation. High conductivity copper has a purity better than 99.9 per cent. This is called cathode copper because the refined copper forms the cathode of an electrolytic cell as shown in Fig. 7.6. The impure copper forms the anode and the electrolyte is an acidulated solution of copper sulphate. On the passage of a direct electric current through the cell, metallic copper from the electrolyte is deposited on the cathode. This upsets the chemical balance of the electrolyte, which then dissolves copper from the anode to restore the balance. The process continues as long as the current flows and the copper deposited upon the cathode has a very high degree of purity. Any impurities are precipitated on the bottom of the electrolytic cell as a dross which has to be removed periodically. Copper produced by this method of refinement is used for electrical conductors and heat exchangers.

For structural purposes, high conductivity copper is insufficiently strong because of its high degree of purity. To increase the strength, impurities in the form of copper oxides are allowed to form in the metal. Such coppers are referred to as *tough pitch* and are used for general purpose sheets, rods and tubes. Tough pitch copper (which has been 'fire-

Table 7.6 Heat treatable, cast aluminium alloys

Type	Composition		Condition	Properties			Applications
	Cu	Other Elements		0.1% P.S. (MPa)	U.T.S. (MPa)	Elong. (%)	
BS 1490:LM4	3.0	Si 5.0 Mn 0.5 Fe 0.8 Ti 0.2	Solution treated at 520 °C for 6 hours. Precipitation hardened at 170 °C for 12 hours	252	294	1	General purpose alloy for sand casting; gravity and pressure die-casting. Withstands moderate stress, shock and hydraulic pressure.
BS 1490:LM8	(Si) (4.5)	Mg 0.5 Mn 0.5 Ti 0.15	Solution treated at 465 °C for 8 hours. Precipitation hardened at 165 °C for 10 hours.	—	280	2	Good casting properties and corrosion resistance. Mechanical properties can be varied by heat treatment.
BS 1490:LM14	4.0	Si 0.3 Mg 1.5 Ni 2.0 Ti 0.2	Solution treated at 510 °C. Precipitation hardened in boiling water for 2 hours.	215	280	—	Pistons and cylinder heads for liquid and air cooled engines. A good general purpose alloy.
BS 1490:LM16	1.2	Si 5.0 Mn 0.5 Ni 0.25	Solution treated at 520 °C for 12 hours; water quenched. Precipitation hardened at 150 °C for 10 hours.	182	231	1	Suitable for castings of intricate shape. High pressure tightness: suitable for valve bodies and cylinder heads. Also used for water jackets and cylinder blocks.

Table 7.7 Heat-treatable, wrought aluminium alloys

Type	Composition		Condition	Properties			Applications
	Cu	Other Elements		0.1% P.S. (MPa)	U.T.S. (MPa)	Elong. (%)	
DTD 372	—	Si 0.5 Mg 0.6 Ti 0.2	Solution treated at 520 °C; quenched. Precipitation hardened at 170 °C for 10 hours.	168	224	18	Good corrosion resistance. Extruded sections such as glazing bars and window sections. Windscreen and sliding roof sections for automobile body-building industry.
BS 1470/7 H114	4.0	Mg 0.8 Si 0.5 Mn 0.7	Solution treated at 480 °C; quenched. Age hardened at room temperature for 4 days.	280	400	10	General purpose alloy suitable for stressed parts in aircraft and other structures. The original 'Duralumin'.
BS 1470/7:H730	—	Mg 1.0 Si 1.0 Mn 0.7	Solution treated at 510 °C; quenched. Precipitation hardened at 175 °C for 10 hours.	150	250	20	Structural members for road and rail vehicles and shipbuilding. Architectural work. Ladders and scaffold tubes. High electrical conductivity: overhead powerlines.
DTD 5074	1.6	Mg 2.5 Zn 6.2 Ti 0.3	Solution treated at 465 °C; quenched. Precipitation hardened at 120 °C for 24 hours.	590	650	11	Strongest commercial alloy. Highly stressed aircraft components. Military equipment.

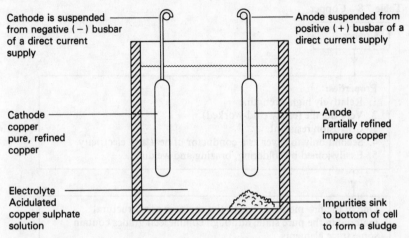

Fig. 7.6 Electrolytic cell for the refinement of copper

refined') does not have such good electrical conductivity or corrosion resistance as electrolytically refined, high purity, cathode copper.

Pure copper is a difficult material to machine to a good surface finish. However the addition of traces of the metal tellurium or the non-metal sulphur produces a *free-cutting copper* with only slightly impaired ductility and conductivity. For example, the addition of 0.5 per cent tellurium forms the chip breaking compound copper-telluride, whilst maintaining an electrical conductivity of 90 per cent that of high conductivity copper. Similarly the addition of 0.4 per cent sulphur produces the chip making compound copper sulphide, whilst maintaining an electrical conductivity of 95 per cent that of high conductivity copper. However the use of sulphur, although cheaper, substantially reduces the strength of the copper. To prevent gassing and porosity during welding, traces of phosphorus are added to the copper. The phosphorus combines with any dissolved oxygen present, and copper treated in this manner is said to be *phosphorus deoxidised*. It should be noted that the presence of only 0.04 per cent phosphorus results in a 25 per cent reduction in electrical conductivity. Therefore phosphorus deoxidised copper is unsuitable for electrical conductors. Table 7.8 shows the relationship and some typical uses of the more commonly available forms of copper, whilst Table 7.9 lists some typical properties.

7.7 High copper content alloys

The first group of copper alloys to be considered are the high copper content alloys, that is, the additional alloying elements represent only a very small percentage of the total. In spite of this, these small additions

164

Table 7.8 Copper

Copper

↓

Properties:
1. Relatively high strength.
2. Very ductile (easily cold-worked).
3. Corrosion resistant.
4. Second only to silver as a conductor of heat and electricity.
5. Easily joined by soldering, brazing and welding.

↓

One of the few metals of use to the engineer as a structural material in the pure state, although commercial grades contain some trace elements.
Availability: Cold drawn rods, bars, wire and tubes. Cold rolled sheet and strip. Extruded sections. Castings. Powder for sintered components.

Cathode copper
Used in the production of copper alloys.

Phosphorus-de-oxidised non-arsenical copper
Welding quality copper. Removal of the dissolved oxygen content prevents gassing and porosity. Used for fabrication, casting, cold impact extrusion and severe presswork.

High conductivity copper
Better than 99.9% pure. Used for electrical conductors and heat exchangers.

Refined tough pitch copper
General purpose copper. Used for roofing, chemical plant, general presswork, decorative metalwork and applications where special properties are not required.

Arsenical tough pitch and phosphorus de-oxidised copper
The addition of arsenic improves the strength at high temperatures. Used for boiler and firebox plates, stays, flue tubes and domestic plumbing.

Table 7.9 Properties of copper

Description	Purity	Oxygen Content	Condition	U.T.S. (MPa)	Elong. (%)	Hardness (H_B)
Electrolytic tough pitch high-conductivity copper	99.90% (min)	0.05%	Annealed	220	50	45
			Hard	400	4	115
Fire refined tough pitch high-conductivity copper	99.85% (min)	0.05%	Annealed	220	50	45
			Hard	400	4	115
Oxygen-free high-conductivity copper	99.05% (min)	—	Annealed	215	60	45
			Hard	340	6	115
Phosphorus de-oxidised copper	99.85% (min)	O_2 nil P 0.013 to 0.05%	Annealed	210	60	145
			Hard	320	4	115
Arsenical copper	99.20% (min)	0.05% O_2 0.3 to 0.5 As	Annealed	220	50	45
			Hard	400	4	115

make a significant change to the properties of the alloy compared with pure copper.

Silver copper

The addition of only 0.1 per cent silver to high conductivity copper raises its annealing temperature by up to 150°C. This is very important where electrical conductors have to be soldered to hard-drawn copper contacts. If pure copper were used, the heat required to make the soldered joint would soften the copper contacts and the increased rate of wear would render them useless, for example the copper segments of an electric motor or generator commutator.

Cadmium copper

Like silver, cadmium has little effect upon the conductivity of the copper. Again, like silver, cadmium also raises the annealing temperature. In addition, however, it also strengthens and toughens the copper, increasing its resistance to metal fatigue. As cadmium copper is substantially free from oxygen, it is not susceptible to 'gassing' when it is braze welded.

Cadmium copper is used for low- and medium-voltage overhead transmission cables where its high conductivity and high strength enables it to be used over relatively long spans. It is also used for traction purposes, e.g. the overhead conductors for electrified railway systems. Because of its resistance to metal fatigue, annealed cadmium copper is also recommended for aircraft wiring where its flexibility is combined with its resistance to metal fatigue caused by vibration.

Chromium copper

This is one of the few non-ferrous alloys which can be heat-treated to improve its mechanical properties. A typical alloy containing 0.5 per cent chromium can be quenched from 100°C. This leaves the alloy in a soft and ductile condition with a rather low electrical conductivity. However, if the metal is reheated to 500°C for approximately two hours, its mechanical and electrical properties are restored.

Since the properties of chromium copper depend upon heat-treatment rather than upon cold-working, it can be used in cast as well as in wrought forms. Similarly, components can be formed from annealed sheets or extruded rod and then hardened after manipulation and machining.

Tellurium copper

This has already been introduced as a *free-cutting* copper alloy in Section 7.6. Tellurium forms stable compounds with copper and the addition of 0.5 per cent makes the copper as machineable as free-cutting brass yet leaves the electrical conductivity relatively unaffected. Tellurium copper has a high corrosion resistance and is used extensively for high-duty electrical contacts in machines and switchgear for use in hostile environments such as mines, ships and chemical plant. The addition of traces of nickel and silicon allows it to be heat treated in a similar manner to chromium copper, but at the expense of some loss of conductivity.

Beryllium copper

Beryllium copper is used where mechanical rather than electrical properties are required. Beryllium copper can be softened by heating it to 800°C and quenching. In this condition the material can be extensively cold-worked and machined. Subsequent heat treatment consists of heating the alloy to between 300°C and 320°C for upwards of two hours. The resulting mechanical properties will depend to some extent upon the degree of cold-working that took place between the first and second treatments.

Beryllium copper is used widely for instrument springs, flexible bellows, corrugated diaphragms (aneroid barometers and altimeters) and the bourdon tubes of pressure gauges.

Hand tools made from beryllium copper are almost as strong and hard wearing as those made from steel but, since they will not strike sparks from other metals or from flint stones, such tools are widely used in hazardous locations where there is a high risk of fire or explosion, for example mines, oil refineries, oil rigs, chemical plant, etc. The high cost of this alloy precludes its use for more conventional applications.

7.8 The brass alloys

These are alloys of copper and zinc. They tend to give poor quality porous castings which depend upon a combination of hot-working and cold-working to consolidate the metal and improve its mechanical properties. The more common brasses are listed in Table 7.10.

The change in properties with change in composition is more readily understood by reference to the copper-zinc phase equilibrium diagram shown in Fig. 7.7. Although more complex than any phase equilibrium diagram considered so far, it is fairly easy to understand if taken section by section. As in previous diagrams the α phase consists of a solid solution, which in this case is a solid solution of zinc in copper. Like all solid solutions it is ductile and suitable for cold-working.

When the amount of zinc present exceeds that required to saturate the α phase solid solution, a β phase is introduced. This new phase is tougher, stronger and harder than the α phase, but considerably less ductile. The β phase can only exist down to 454°C below which it is modified to β' phase. This modified phase lowers the ductility and malleability still further. If even more zinc is added an even more brittle and less ductile phase is introduced called γ brass. Commercial brasses usually contain only α or $\alpha + \beta'$ phases. The effect of these phases on the properties of a range of brasses is shown in Fig. 7.8, and these properties should be compared with the applications listed in Table 7.10.

The $\alpha + \beta$ brasses are largely used for hot-working (hot-stamping and extrusion) above 454°C. This is because the β phase is much more malleable and ductile than the β' phase which exists at room temperature. The addition of lead or of tellurium to $\alpha + \beta'$ (duplex) brasses gives them free-cutting properties so that they machine easily. These additional alloying elements have little effect upon the strength, hardness and ductility of the brass.

7.9 Tin-bronze alloys

These are alloys of copper and tin together with a deoxidiser. The deoxidiser is essential to prevent the tin content oxidising during casting or hot-working. Oxidation of the tin would result in a weak, 'scratchy' bronze. Two deoxidisers are commonly used.

(a) *Phosphorus* in the 'phosphor-bronze' alloys.
(b) *Zinc* in the 'gun-metal' alloys.

Some typical tin-bronze alloys are listed in Table 7.11.

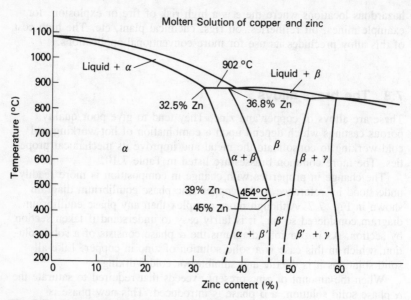

Fig. 7.7 Copper-zinc phase equilibrium diagram

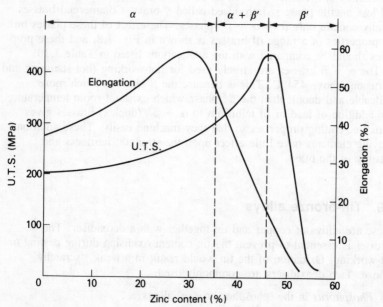

Fig. 7.8 Effect of composition on the properties of brass

Table 7.10 Brass alloys

Name	Composition (%)			Applications
	Copper	Zinc	Other elements	
Cartridge brass	70	30	—	Most ductile of the copper-zinc alloys. Widely used in sheet metal pressing for severe deep drawing operations. Originally developed for making cartridge cases, hence its name
Standard brass	65	35	—	Cheaper than cartridge brass and rather less ductile. Suitable for most engineering processes
Basis brass	63	37	—	The cheapest of the cold working brasses. It lacks ductility and is only capable of withstanding simple forming operations
Muntz metal	60	40	—	Not suitable for cold-working but hot-works well. Relatively cheap due to its high zinc content, it is widely used for extrusion and hot-stamping processes
Free-cutting brass	58	39	3% lead	Not suitable for cold-working but excellent for hot-working and high-speed machining of low strength components
Admiralty brass	70	29	1% tin	This is virtually cartridge brass plus a little tin to prevent corrosion in the presence of salt water
Naval brass	62	37	1% tin	This is virtually Muntz metal plus a little tin to prevent corrosion in the presence of salt water

Table 7.11 Tin-bronze alloys

Name	Composition (%)					Application
	Copper	Zinc	Phosphorus	Tin		
Low-tin bronze	96	—	0.1 to 0.25	3.9 to 3.75		This alloy can be severely cold-worked to harden it so that it can be used for springs where good elastic properties must be combined with corrosion resistance, fatigue resistance and electrical conductivity, e.g. contact blades
Drawn-phosphor-bronze	94	—	0.1 to 0.5	5.9 to 5.5		This alloy is used in the work-hardened condition for turned components requiring strength and corrosion resistance, such as valve spindles
Cast phosphor-bronze	Rem.	—	0.03 to 0.25	10		Usually cast into rods and tubes for making bearing bushes and worm wheels. It has excellent anti-friction properties
Admiralty gunmetal	88	2	—	10		This alloy is suitable for sand casting where fine-grained, pressure-tight components such as pump and valve bodies are required
Leaded-gunmetal (free-cutting)	85	5 (5% lead)	—	5		Also known as 'red brass', this alloy is used for the same purposes as standard, admiralty gunmetal. It is rather less strong but has improved pressure tightness and machining properties
Leaded (plastic) bronze	74	(24% lead)	—	2		This alloy is used for lightly loaded bearings where alignment is difficult. Due to its softness, bearings made from this alloy 'bed in' easily

Unlike the brasses which are largely used in the wrought condition (rod, sheet, stampings, etc.), only low-tin content bronze alloys can be worked and most bronze components are in the form of castings. Tin-bronze alloys are more expensive than brass alloys, but they are stronger, more corrosion and wear resistant, and give sound, pressure-tight castings which are widely used for steam and hydraulic valve bodies and mechanisms.

The phase equilibrium diagram for copper-tin alloys is shown in Fig. 7.9. This diagram is too complex to warrant detailed study at this stage, but it should be noted that up to about 10 per cent tin (depending upon temperature) only the α phase solid solution of tin in copper is present. Thus, as for all solid solutions, these low-tin content bronze alloys are ductile and can be cold-worked. As previously stated, care must be taken to prevent oxidation as this will weaken the alloy and make it 'scratchy' and brittle. In these low tin-bronze alloys the deoxidising agent is always phosphorus and the resulting alloy is a *phosphor bronze*. After the alloy has become work-hardened due to cold-working (cold-rolling, wire-drawing, etc.), the resulting strip or wire can be formed into instrument springs, electrical contacts, and similar components.

If the tin content is increased beyond about 13 to 14 per cent, the alloy will show considerable amounts of the δ phase at room temperature. These duplex alloys of $\alpha + \delta$ phase are too brittle to be cold-worked and are used solely for the production of castings. These consist of fine particles of the hard δ phase dispersed throughout a matrix of tough, ductile

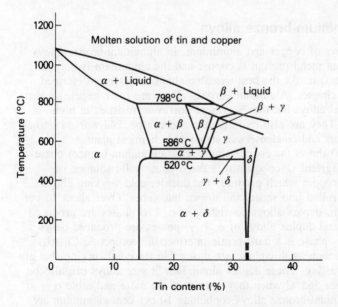

Fig. 7.9 Copper-tin phase equilibrium diagram (copper-rich alloys)

α phase. This results in a low-friction, hard-wearing surface, and a sound pressure-tight casting free from porosity. These casting alloys may be deoxidised with traces of phosphorus so as to leave not more than a residue of 0.04 per cent in the finished casting. Alternatively the phosphorus content may be increased up to 1.0 per cent when increased fluidity, strength and corrosion resistance is required. Increasing the phosphorus content to this level results in a reduction in toughness and a corresponding increase in brittleness, together with an improvement in anti-friction properties (see Chapter 9).

Zinc may be used in casting alloys as a deoxidiser and, as it replaces a substantial amount of the tin content, it reduces the cost of the alloy which is an important factor with large castings. As has already been stated alloys containing zinc as a deoxidiser are referred to as 'gun-metal'. Such alloys give excellent pressure-tight, corrosion-resistant castings but they do not have the anti-friction properties of the more expensive phosphor-bronzes. Gun-metals are used for such applications as hydraulic and steam valve body castings and castings for marine use. The alloy gets its name from the fact that it was originally used for casting cannon barrels for the Admiralty.

The more important phases found in the copper-tin alloy system may be summarised as follows:

α phase → tin (Sn) in solution in excess copper (Cu)
β phase → copper (Cu) in solution in excess tin (Sn)
δ phase → a hard, brittle intermetallic compound $Cu_{31}Sn_8$

7.10 Aluminium-bronze alloys

These are alloys of copper and aluminium. In aluminium-bronze alloys the predominant metal present is copper and the aluminium is only the alloying element, unlike the heat-treatable aluminium alloys described earlier in this chapter. Aluminium-bronze alloys are more expensive than the 'tin-bronze' alloys but they are more corrosion resistant at high temperatures. They are also more ductile and can be cold-worked into tubes for boilers and condensers in steam and chemical plant.

Figure 7.10 shows the copper-aluminium (aluminium-bronze) phase equilibrium diagram. Once again it is the α phase solid solution of aluminium in copper which produces the ductile, cold-working alloys which can be rolled into sheets and drawn into tubes. Over about 10 per cent aluminium, duplex alloys consisting of α + β phases are produced above 565°C and duplex alloys of α + $γ_2$ phases are produced below 565°C. The $γ_2$ phase is a hard brittle intermetallic compound Cu_9Al_4. Thus alloys containing this phase are unsuitable for cold-working, but are excellent for casting. These duplex aluminium bronze alloys can also be hot-worked over 565°C when they consist of the more malleable α + β phases. Aluminium-bronze alloys containing 10 per cent aluminium are one of the few non-ferrous alloys which can be heat-treated like steel.

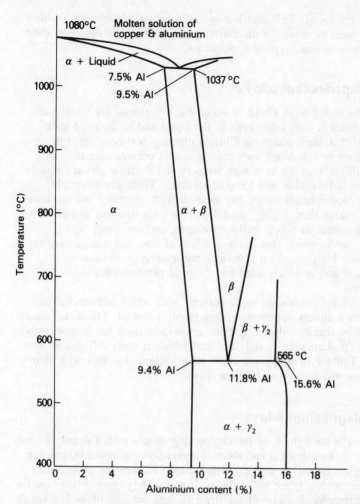

Fig. 7.10 Copper-aluminium phase equilibrium diagram (copper-rich alloys)

For example, quenching this alloy from 900°C produces a structure which consists entirely of the β' phase. This phase is analogous with martensite in steel since it is very hard and brittle and has acicular (needle-like) crystals. To increase the toughness of the alloy with some small loss of hardness, like steel, it can be tempered by reheating it to about 400°C and quenching it. Further, slow cooling under equilibrium conditions from above 565°C anneals the alloy and restores the normal α + γ_2 phases at room temperature.

Aluminium bronzes are highly corrosion resistant particularly at high temperatures since a homogeneous film of aluminium oxide forms on the surface of the metal. This film is also 'self-healing' if damaged. Table 7.12 lists some examples of aluminium bronze alloys together with their composition and some typical applications.

7.11 Cupro-nickel alloys

Copper and nickel form alloys in which the two metals are wholly soluble (miscible) in each other both in the liquid and in the solid state. Because of this, their phase equilibrium diagram was considered in some depth in Section 3.3. Since only the α phase is present over the entire range of alloys, they all have high strength and ductility and are equally suitable for hot-working and for cold-working. These are relatively expensive copper-based alloys, but their strength, ductility and corrosion resistance makes them highly suitable for such applications as high duty boiler and condenser tubes, bullet envelopes, and resistance wires.

Monel metal, which also contains traces of iron and manganese, has exceptionally high corrosion resistance particularly at elevated temperatures and is widely used for chemical plant and marine applications.

Nickel silvers contain up to 20 per cent zinc which reduces the cost and imparts a silvery appearance when highly polished. The nickel-silver alloys can be readily cold-worked and are widely used for domestic table cutlery as an alternative to stainless steel which is more difficult to work to shape. Table 7.13 lists some cupro-nickel alloys, together with their composition and some typical applications.

7.12 Magnesium alloys

Magnesium is the lightest of the engineering metals with a density of only 1.7 g mm^{-3}. Its electrical and thermal conductivity is about 60 per cent that of high conductivity copper. It has a high affinity for oxygen and burns in air with a fierce white flame. It was widely used at one time for flash photography. Because of the fire risk, and because of its low tensile strength, magnesium is only used as an alloying element or as the basis of a range of ultra-lightweight alloys.

Although not as strong as aluminium alloys, magnesium alloys have a much lower density, so that in many instances their strength-to-weight ratio can actually be superior. The magnesium alloys fall into two categories:

(*a*) casting alloys;
(*b*) wrought alloys.

Magnesium alloys contain aluminium, zinc, zirconium, and manganese, together with 'rare-earth' metallic elements in some instances. Like the aluminium alloys, magnesium alloys can be annealed by solution treatment. The alloy is heated at 380°C for eight hours, when the temperature

Table 7.12 Copper-aluminium (aluminium bronze) alloys

Composition (%)				Condition	Properties				Applications
Al	Fe	Cu	Other Elements		0.1% P.S. (MPa)	U.T.S. (MPa)	Elong. (%)	Hardness (H_D)	
5	—	Rem.	Mn or Ni Up to 4%	Annealed	112	350	70	80	Cold-worked for decorative purposes such as imitation jewellery. Tubes for engineering applications, resistant to corrosion and oxidation.
				Hard	532	700	4	200	
8	—	Rem.	Fe, Mn, Ni Up to 2%	Hot-worked	140	392	45	100	Chemical engineering plant suitable for use at moderately elevated temperatures.
10	5	80	Ni–5%	Hot-forged	420	658	20	215	General engineering forgings combining strength and corrosion resistance – can be heat treated.
9.5	2.5	Rem.	Ni and Mn Up to 1.0%	As cast	168	476	30	115	General purpose alloy for both die-casting and sand-casting.
12	—	Rem.	Fe, Mn, Ni from 5% to 8%	As cast	434	504	3	250	A hard, rigid alloy containing the γ_2 phase. Used where heavy compressive loads are involved. Good wear resistance.

Table 7.13 Cupro-nickel alloys

Composition (%)				Properties					Applications
Cu	Ni	Other Elements	Condition	0.1% P.S. (MPa)	U.T.S. (MPa)	Elong. (%)	Hardness (H_D)		
80	20	—	Annealed	98	308	45	75		Very high ductility and corrosion resistance, will withstand severe cold working.
			Hard	420	490	5	165		
70	30	—	Annealed	98	322	45	80		Used for condenser and heat-exchanger tubes where high corrosion resistance is required.
			Hard	490	588	5	175		
60	40	—	Annealed	—	350	45	90		'Constantan' electrical resistance wire – high specific resistance and low temperature coefficient. Also used in thermocouples.
			Hard	—	588	5	190		
29	68	Fe 1.25 Mn 1.25	Annealed	196	490	45	120		'Monel Metal'. Good mechanical properties combined with excellent corrosion resistance properties. Used for chemical plant.
			Hard	518	658	70	220		

is raised to 410°C for a further sixteen hours. Magnesium alloys can be precipitation hardened by heating at 190°C for some ten to twelve hours.

The melting and casting of magnesium alloys is difficult and dangerous in view of the flammability of magnesium if accidentally overheated. These alloys are generally melted under a flux containing calcium and sodium fluorides to exclude atmospheric oxygen. As the molten metal is poured into the mould it is dusted with flower of sulphur. The sulphur burns on contact with the hot metal and blankets it with sulphur dioxide gas which excludes atmospheric oxygen without which the molten alloy cannot ignite. The presence of sulphur dioxide gas makes efficient fume extraction and good ventilation most important. Table 7.14 lists some cast and wrought magnesium alloys together with their composition, properties and some typical applications.

7.13 Zinc alloys

Pure zinc is an interesting metal. Its boiling point is so low that it is the only commercial metal which can be refined by distillation. It has a density of 7.1 g mm^{-3} and a melting point of only 420°C. The pure metal is relatively weak, but is widely used as a coating on steel to prevent corrosion by atmospheric attack and zinc-coated low-carbon steel sheet is known as 'galvanised iron' (see Section 13.8).

Zinc-based alloys are used almost entirely for pressure die-casting. They are widely used for such components as car door handles, carburettor and fuel pump bodies, and other lightly stressed components. Zinc pressure die-castings can be machined, but they cannot be soldered or welded. The popularity of zinc-based alloys for die-casting is due to their:

(a) high fluidity which enables complex castings with thin sections to be made;
(b) low melting point which reduces die wear;
(c) narrow freezing range which results in rapid solidification and allows high rates of production;
(d) easy polishing and electro-plating properties;
(e) adequate strength for small components.

A typical zinc based alloy marketed under the trade name of 'Mazak' contains:

Aluminium 4%
Copper   2.7%
Zinc   remainder.

The zinc must be better than 99.9 per cent pure, otherwise even minute traces of impurities such as cadmium, tin and lead will lead to intercrystalline brittleness, swelling, corrosion and pitting of the electro-plated surface. It was the lack of high-purity zinc which gave zinc alloy die-castings their poor reputation for quality in the early days of the process.

Table 7.14 Magnesium alloys

Composition (%) (Remainder Mg)						Condition	Properties			Applications
Al	Mn	Zn	Zr	Th	Rare earths		0.1% P.S. (MPa)	U.T.S. (MPa)	Elong. (%)	
10.0	0.3	0.7	—	—	—	Chill-cast	115	200	2	Light-weight castings for the aircraft and high-performance car industry. For example: Landing wheels, road wheels, crank cases, and miscellaneous coatings.
—	—	4.0	0.7	—	1.2	As Cast	95	170	5	
						Heat-treated	130	215	4	
—	—	—	0.7	3.0	—	Heat-treated	100	210	8	
—	1.5	—	—	—	—	Rolled	95	200	5	Petrol tanks, oil tanks and other lightly stressed sheet metal components. Light-weight sections.
—	1.0	—	—	3.0	—	Rolled	215	280	10	
6.0	0.3	1.0	—	—	—	Forged	155	280	8	Light-weight forgings for the aircraft industry such as airscrew blades and undercarriage components.
						Extruded	140	215	8	
—	—	3.0	0.7	—	—	Forged	170	265	8	
						Extruded	215	310	8	

Casting alloys — rows 1–3

Wrought alloys — rows 4–7

Zinc pressure die-castings should have a strength of about 320 MPa and a high rigidity.

The corrosion resistance of zinc-based die-castings can be increased by the 'chromate passivation' process. The castings are cleaned and immersed in a solution of dilute sulphuric acid and sodium bichromate for not more than one minute. They are then washed and dried. The resultant passive film prevents corrosion in damp atmospheric conditions when the die-casting has not been electro-plated.

The range of alloys discussed in this chapter is only intended as a representation of the range available. Any one group could be a study in its own right. The soft solders (tin-lead alloys) were discussed in Section 3.10, and bearing metals will be discussed in Chapter 9.

8 Polymeric materials

8.1 Additives

The principles underlying the structure and behaviour of polymeric materials were considered in Chapter 2. This chapter will now consider the properties and uses of these materials in greater depth.

It is not usual to use pure resin to mould a plastic product. The appearance and performance of most plastic and elastic polymers can be improved by the use of various additives which will now be considered.

Plasticisers

These are added to polymeric materials to reduce their rigidity and brittleness and improve their flow properties whilst being formed or moulded. There are two main groups of plasticisers.

(a) *Primary plasticisers* are used to neutralise partially the Van der Waal's forces between adjacent molecular chains by introducing monomers whose polar groups neutralise those of the polymer groups and allow greater mobility between adjacent polymer chains.

(b) *Secondary plasticisers* are monomers of a compatible but inert material without polar groups which may be added to provide mechanical separation of the polymer chains in the same way that a lubricant separates a shaft from its bearing. Separation of the polymer chains in this manner reduces the Van de Waals forces of attraction between them, as shown in Fig. 8.1. Secondary plasticisers may, themselves, be divided into two groups according to the method of application.

X = Strong intermolecular (Van der Waal's) forces

(a)

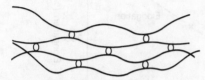

O = Secondary plasticiser separating
the polymer chains and weakening
the Van der Waal's forces

(b)

Fig. 8.1 Secondary plasticiser (*a*) Without plasticiser (*b*) With plasticiser

(i) *Internal plasticisation.* Small amounts of plasticiser are added
during polymerisation. The additive produces bulky side groups
which achieve separation of the polymer chains as they form
during the polymerisation process (secondary plasticisation). For
example polyvinyl chloride (PVC) which is a rigid rather brittle
plastic material, can be made flexible for raincoats, garden
hosepipes and the insulation of electric cables by the addition of
15 per cent vinyl acetate as a secondary plasticiser during
polymerisation.

(ii) *External plasticisation.* This is the more common method of
plasticisation. The plasticiser, in the form of a low-volatility
liquid solvent, is added after polymerisation. it disperses
throughout the plastic, filling the voids between the polymer
chains and acting as a lubricant. Again, this is secondary
plasticisation as the plasticiser separates the polymer chains and
weakens the Van der Waal's forces as previously explained. Only
the amorphous zones are lubricated in this manner as highly
crystalline polymers will not absorb sufficient solvent plasticiser.
Although it is difficult, effectively, to plasticise high crystallinity
polymers, the presence of a plasticiser helps to reduce the
amount of crystallinity present by interfering with the formation
of orderly arrays in the polymer chains. By reducing the amount
of crystallinity present, the plasticiser also reduces the glass
transition temperature (Tg). Figure 8.2 shows the effect of
adding a plasticiser to polyvinyl chloride (PVC).

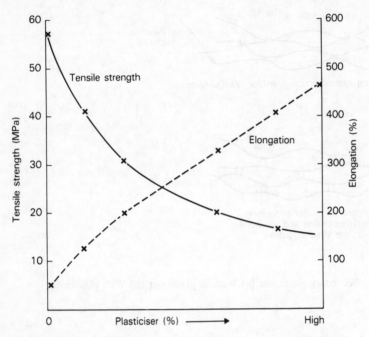

Fig. 8.2 Effect of a plasticiser on the mechanical properties of PVC

Fillers

These have a considerable influence on the properties of mouldings produced for any given polymeric material. They improve the impact strength and reduce shrinkage during moulding. Typical fillers together with their properties are listed in Table 8.1.

Fillers are essential in thermosetting moulding powders and may be present in quantities up to 80 per cent by weight. They are less frequently found in thermoplastics where their presence is normally limited to 25 per cent by weight. The exceptions are thermoplastic floor tiles which may contain up to 40 per cent calcium carbonate by weight.

The selection of a filler is usually determined by the properties it can impart to the plastic product and also its cost. However, all fillers must have a low moisture absorption rate; they must not adversely affect the colour or surface finish of the product; they must not cause abrasive wear in the processing equipment; and they must be capable of being 'wetted' by the resin. Figure 8.3 shows the effect of various filler materials on the stress/strain curve for a typical phenolic resin, whilst Fig. 8.4 shows the effect of various filler materials on the elastic modulus (rigidity) of a typical polyester resin.

Stabilisers

Stabilisers are used to prevent the degradation of polymeric materials occuring when they are exposed to heat, sunlight and weathering. Such

Table 8.1 Fillers

Filler material	Properties
Glass fibre:	good electrical insulation properties
Wood flour: Calcium Carbonate:	low cost, high bulk, low strength.
Asbestos:	heat resistant (no longer recommended; health hazard).
Aluminium powder:	high mechanical strength.
Shredded paper: Shredded cloth: Mica granules:	good strength, combined with reasonable electrical insulation properties.

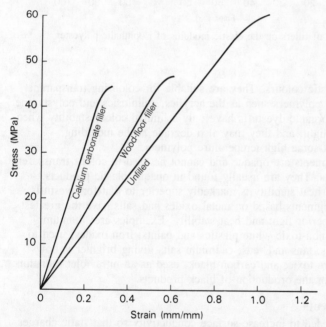

Fig. 8.3 Effect of filters on the mechanical properties of a phenolic thermoset

degradation is usually accompanied by colour change, deterioration in mechanical properties, cracking and surface crazing. Since these are environmental changes, they will be dealt with more fully in Section 13.14.

Colourants

These can be subdivided into dyestuffs, organic pigments, and inorganic pigments. Dyestuffs are usually aromatic organic chemicals which are soluble in a variety of solvents. They absorb light selectively to produce

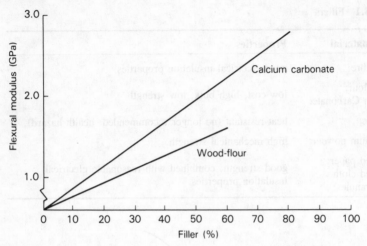

Fig. 8.4 Effect of fillers on the elastic modulus of isophthalic polyester

their characteristic colours. They are suitable for colouring transparent and translucent polymers such as the acrylics, cellulosics and polystyrene. Unfortunately organic dyestuffs have only a limited colour stability when exposed to sunlight, and they may also degrade at the moulding temperatures of some high temperature polymers.

Organic pigments are opaque and cannot be used to colour transparent plastic materials. They are usually found in opaque plastic products. Their light and heat stability is markedly superior to that for dyestuffs.

Inorganic pigments based on metal oxides and salts have the greatest opacity and superior light and heat stability. Examples are: titanium oxide, used in non-toxic white plastics and paints; iron oxides, used to provide yellows, tans and reds; cadmium salts giving brighter yellows and reds but are toxic; and carbon black, used as an ultraviolet radiation absorber and for the production of black products.

Antistatic agents

These are included to increase surface conductivity so that static charges can leak away. This prevents the attraction of dust particles and reduces the risk of explosion in hazardous environments caused by the spark associated with an electrical discharge. It also prevents electric shocks which can occur when plastic materials are handled in very dry climates.

8.2 General properties of polymeric materials

The properties of polymeric materials can vary widely, but they all have the following properties in common.

Electrical insulation

All polymeric materials exhibit good electrical insulation properties. However, their usefulness in this field is limited by their low heat resistance and their softness. Thus they are useless as formers on which to wind electric radiator elements, and as insulators for use out of doors where their relatively soft surface would soon be roughened by the weather. Dirt collecting on this roughened surface would then provide a conductive path, causing a short circuit.

Strength/weight ratio

Polymeric materials vary in strength considerably. Some of the stronger, such as nylon, compare favourably with the weaker metals. All polymeric materials are much lighter than any of the metals used for engineering purposes. Therefore, properly chosen and proportioned, their strength/weight ratio compares favourably with many light alloys and they are steadily taking over engineering duties which, until recently, were considered the prerogative of metal.

Corrosion resistance

All polymeric materials are inert to most inorganic chemicals and can be used in environments which are hostile even to the most corrosion-resistant metals. The synthetic rubbers which are a product of polymer chemistry are superior to natural rubber (polyisoprene) since they are not attacked by oils and greases. The degradation of polymers will be considered more fully in Section 13.14.

8.3 Properties and applications of elastomers

Elastomers are substances which permit extreme reversible extensions to take place at normal temperatures. Natural rubber is an obvious and very important elastomer. Because elastomers are cross-linked like thermosets, their polymer chains cannot slide over each other under applied loads to take on a permanent set. However, unlike true thermosets which have very many cross-links, elastomers have relatively few cross-links. Further, the helical molecular chain of an elastomer uncoils when stressed as shown in Fig. 8.5, and recovers when the load is removed in a similar manner to a coil spring.

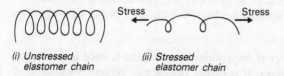

(i) Unstressed elastomer chain   (ii) Stressed elastomer chain

Fig. 8.5 Behaviour of an elastomer under stress

Many components do not require great strength but they do require softness, flexibility and reversible elongation. Thus the rubbers (natural and synthetic elastomers) are ideal for such applications as resilient floor coverings, weatherstripping, footwear, vehicle tyres, joint sealants, expansion joints, and anti-vibration mountings. Like other polymers, elastomers become brittle below their glass transition temperatures. The Tg for most synthetic rubbers is about $-20°C$ and for silicone and natural rubber it is about $-60°C$.

The elastomers are usually addition polymerised as thermoplastics and then cross-linked (vulcanised) with sulphur at approximately every five-hundredth carbon atom. Increased vulcanisation increases the cross-linking and this, in turn, increases the stiffness and reduces the elongation percentage of the material. For example, fully vulcanised, natural rubber becomes a rigid, brittle thermoset called ebonite.

The uses to which elastomers (rubbers) may be put by engineers may be classified as follows

Vibration insulation and isolation
(a) shock absorbers
(b) anti-vibration machine and engine mountings
(c) sound insulation.

Distortional systems
(a) correctives for misalignment: such as flexible couplings
(b) changing shapes: such as belts, flexible hose, covered rollers, and tyres
(c) seals and gaskets
(d) rubber hydraulics (forming tools).

Protective systems
(a) protection against abrasion
(b) protection against corrosion
(c) electrical insulation
(d) protective clothing: gloves, aprons, boots.

Some typical groups of elastomers used in engineering are as follows.

Acrylic rubbers. These are derived from the same family of polymeric materials as 'Perspex' but, in order to give them the properties associated with elastomers, they are not so heavily cross-linked. This group of rubbers have excellent resistance to oils, oxygen, ozone, and ultraviolet radiation, and they are used as the basis for the latex paints developed for motor vehicles.

Butyl rubber. This rubber is impervious to gases and is used as a vapour barrier and for hose linings. It is highly resistant to outdoor weathering and ultraviolet radiation and is used for construction industry sealants.

Nitrile rubber. This has excellent resistance to oils and solvents and can be readily bonded to metals. It is used for petrol and fuel oil hoses, hose linings, and aircraft fuel tank linings. It is also resistant to refrigerant gases.

Polychloroprene rubber (neoprene). This was the original synthetic rubber developed during World War 2. It has good resistance to oxidation, ageing, and weathering. It is resistant to oils, solvents, abrasion and elevated temperatures. Because of its chlorine content it is fire resistant. It is used as a flexible electrical insulator and for gaskets, hoses, engine mounts, sealants, rubber cements and protective clothing.

Polyisoprene (natural rubber). This is derived from the sap of a tree called *Hevea Brasiliensis*. It has a low hysteresis, a relatively high tensile strength, and a very low glass transition temperature. Unfortunately it is readily attacked by solvents, petrol, mineral oils, and ozone. It degrades (perishes) rapidly in the presence of strong sunlight.

Modified by additives and vulcanisation to give it increased strength and wear-resistant properties, natural rubber is used for vehicle tyres as it has excellent anti-skid properties.

Polysulphide rubber (thiokol). Although this rubber has low mechanical strength, its resistance to solvents and its impermeability to gases is excellent and its weathering characteristics are outstanding. It also has good bonding properties and is widely used in the construction industries as a sealant. Thiokol and polyurethane rubbers are also used as fuels for solid-fuel rockets.

Polyurethane rubber. Polyurethane can be formulated to give either plastic or elastic properties. Although it has relatively high strength and abrasion resistance, it is of little use for vehicle tyres as it has low skid resistance. However, its outstanding service life makes it suitable for cushion tyres for warehouse trucks and forklift trucks where low speeds and dry floor conditions make its low skid resistance more acceptable. It is also used for shoe heels, painting rollers, mallet heads, oil seals, diaphragms, anti-vibration mountings, gears, and pump impellers.

Rubber hydrochloride. This material is better known as 'Pliofilm' and is used to form a transparent film for the vacuum packaging of foods and DIY hardware. It is easily identified by its unusually high tensile strength and tear resistance.

Silicone rubbers. Although silicone rubber has a relatively low tensile strength, it has an exceptionally wide working temperature range of $-80°C$ to $+235°C$. Thus it often outperforms other rubbers which are superior at room temperature but which cannot exist at such temperature extremes. It is also used in space vehicles and artificial satellites.

Styrene-butadiene rubber (SBR). A general-purpose synthetic rubber used for tyres, floor-tiles, and latex paints. It is superior to natural rubber in respect of skid resistance, solvent resistance, and weathering.

8.4 Properties and applications of typical themoplastics

Polyethylene (polythene)

This is one of the most versatile and widely used plastic materials. It remains tough and flexible over a wide range of temperatures and has good dimensional stability. It is easily moulded and is used in a wide range of domestic goods such as buckets, bowls, food containers, bags and squeezy bottles. It is also used commercially for water piping, chemical equipment and for electrical insulation. It is resistant to most solvents and has good weathering properties. Unfortunately it degrades when exposed to strong sunlight unless it contains a UV filter pigment such as carbon black. Polyethylene can exist in the amorphous state or with varying degrees of crystallinity, low density polyethylene has a branched molecular chain which makes the formation of crystallites difficult. The structure of the branched molecular chain is shown in Fig. 8.6(*a*). High density polyethylene has a simple linear chain and this lends itself to the close packing essential for high density and for the orderly structure required to form crystallites. The structure of a simple linear chain of high density polyethylene is shown in Fig. 8.6(*b*). Typical properties are:

	Low density	*High density*
Crystallinity (%)	60	95
Density (kg m^{-3})	920	950
Tensile strength (MPa)	11.0	31.0
Elongation (%)	100 to 600	50 to 800
Impact value (J)	No fracture	5 to 15
Tg (°C)	−120	−120
Tm (°C)	115	138
Maximum service temperature (°C)	85	125

Polypropylene

This is a tough, rigid, lightweight material with similar properties to polythene but with better heat resistance. It has good mechanical properties, and good resistance to attack by acids, alkalis and salts even at high temperatures. It is widely used for chemical plant equipment and domestic hardware, and is also used for moulding hospital and laboratory equipment. It can be drawn into fibres for rope and net making, and it can be rolled into sheets for packaging. It is a good moulding material for electrical insulators. Typical properties are:

Crystallinity (%)	60
Density (kg m^{-3})	900

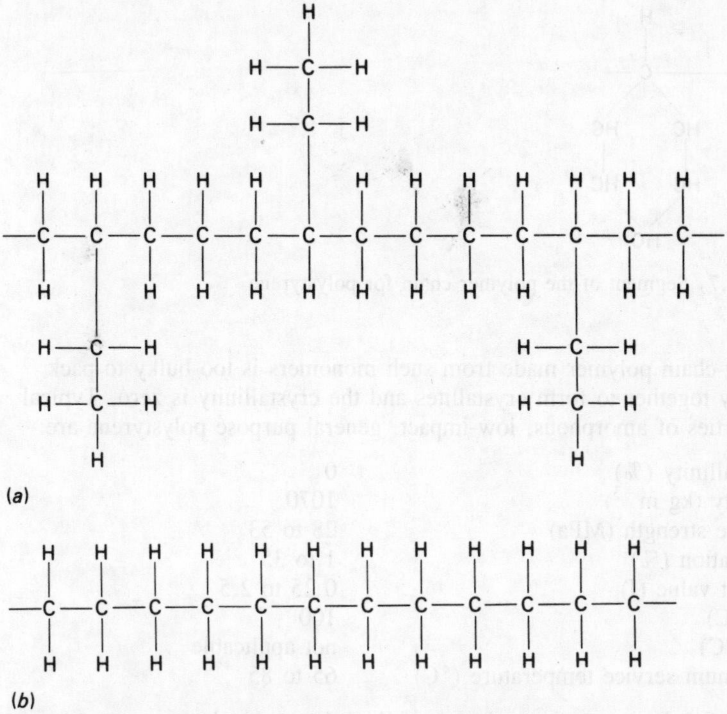

Fig. 8.6 Polymer chains for polyethylene (a) Polymer chain for low-density (branched) polyethylene (b) Polymer chain for high-density (linear) polyethylene

Tensile strength (MPa)	30 to 35
Elongation (%)	50 to 600
Impact value (J)	1 to 10
Tg (°C)	−25
Tm (°C)	176
Maximum service temperature (°C)	150

Polystyrene

This is a tough dense plastic which is hard and rigid, and has good dimensional stability. It can be moulded to give a high surface gloss and is used for such articles as domestic holloware, and refrigerator trays. Although it has good mechanical properties even at low temperatures, it tends to be brittle because of its high glass transition temperature. Being an amorphous material it has no clearly defined melting temperature. Unfortunately it is attacked by petrol and other organic solvents. It can be foamed, and rigid-foamed polystyrene is used for heat and sound insulation blocks in building and refrigeration. it is also used for ceiling tiles, moulded packaging, and for buoyancy aids. Polystyrene has an aromatic ring as a side branch to the monomer and this is shown in Fig. 8.7. Thus

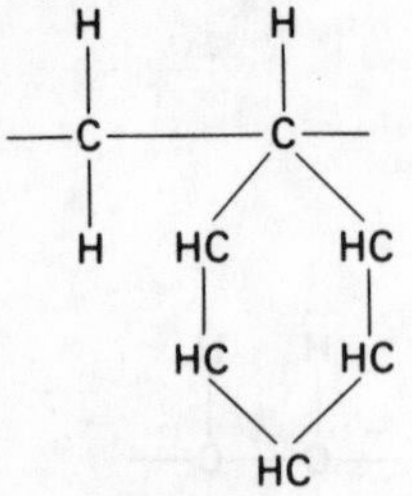

Fig. 8.7 Segment of the polymer chain for polystyrene

a long-chain polymer made from such monomers is too bulky to pack closely together to form crystallites and the crystallinity is zero. Typical properties of amorphous, low-impact, general purpose polystyrene are:

Crystallinity (%)	0
Density (kg m^{-3})	1070
Tensile strength (MPa)	28 to 53
Elongation (%)	1 to 35
Impact value (J)	0.25 to 2.5
Tg (°C)	100
Tm (°C)	not applicable
Maximum service temperature (°C)	65 to 85

(*Note:* for the properties and uses of 'high-impact' polystyrene, see 'ABS', later in this section.)

Polyvinyl chloride (PVC)

Unplasticised polyvinyl chloride is hard and tough but it can be moulded into a variety of hard-wearing holloware articles such as buckets and plant pots. It is also used for builders' hardware such as rain guttering, downpipes and drainpipes. To form and mould rigid PVC it has to be heated above its Tg of 87°C to make it soft and flexible. Upon cooling below its Tg, it becomes rigid again. When plasticised it becomes flexible and rubbery, in which condition it can be used for waterproof clothing, hose pipes, electric cable insulation, and chemical tank linings. It offers good resistance to attack by water, acids, alkalis and most common solvents. Unfortunately it hardens and becomes brittle with age. Typical properties are:

	Unplasticised	Plasticised
Crystallinity (%)	0	0
Density (kg m^{-3})	1400	1300
Tensile strength (MPa)	49	7 to 25
Elongation (%)	10 to 130	240 to 380
Impact value (J)	1.5 to 18.0	not applicable

Tg (°C)	87	87
Tm (°C)	not applicable	not applicable
Maximum service temperature (°C)	70	60 to 105

Polytetrafluoroethylene (PTFE)

This is one of the most versatile and important plastics despite its relatively high cost. It is suitable for the manufacture of tough mouldings and as a non-stick, anti-friction coating (Teflon). It does not burn, neither is it attacked by any known reagent or solvent. It is a good electrical insulator and it has the lowest coefficient of friction of any known solid. It is widely used for bearings, fuel hoses, gaskets and tapes, and as a non-stick coating for cooking utensils. It is also used as a lining for chemical equipment because of its resistance to chemical attack. Typical properties are:

Crystallinity (%)	90
Density (kg m^{-3})	2170
Tensile strength (MPa)	17 to 25
Elongation (%)	200 to 600
Impact value (J)	3 to 5
Tg (°C)	-126
Tm (°C)	327
Maximum service temperature (°C)	260

Polymethyl methacrylate (Perspex)

This is a relatively lightweight material with excellent optical properties. It is also strong and rigid with excellent electrical insulating properties. Unfortunately it is easily scratched, and it softens in boiling water. Also it is attacked by petrol and many organic solvents. It is used for such diverse purposes as lenses, dentures, aircraft windows, sinks, baths, lighting fittings and diffusers, advertising displays, and display lighting. Typical properties are:

Crystallinity (%)	0
Density (kg m^{-3})	1180
Tensile strength (MPa)	50 to 70
Elongation (%)	3 to 8
Impact value (J)	0.5 to 0.7
Tg (°C)	0
Tm (°C)	not applicable
Maximum service temperature (°C)	95

Acrylonitrile-butadiene-styrene (ABS)

This is a tough, strong material with high impact resistance, hence it is often referred to as 'high-impact' polystyrene. It is widely used for moulding television and radio set cabinets, motor vehicle radiator grills and panels, battery cases, crash helmets, water pump and refrigerator

192

parts. It has high dimensional stability and remains tough at sub-zero temperatures. It also resists attack by acids, alkalis, and most petroleum derivatives. High-impact polystyrene is made by co-polymerising low-impact polystyrene with about 5 per cent butadiene rubber. An even tougher variant can be produced by co-polymerising the high impact polystyrene a stage further with acrylonitrile, as shown in Fig. 8.8, to form acrylonitrile-butadiene-styrene (ABS). Typical properties are:

Crystallinity (%)	0
Density (kg m^{-3})	1100
Tensile strength (MPa)	30 to 35
Elongation (%)	10 to 40
Impact value (J)	7 to 12
Tg (°C)	−55
Tm (°C)	not applicable
Maximum service temperature	100

Polyamides (nylon)

Polyamides are that group of polymeric materials known as 'nylons' and were the first of the 'engineering' or high strength thermoplastics. Polyamides are produced by the reaction of diamine with an organic acid. One of the most commonly used nylons is produced by reacting hexa-methylenediamine with adipic acid to give hexamethylenedipamide (com-monly abbreviated 'polyamide'). This is referred to as nylon 6.6 ('66'), and this notation indicates that there are six carbon atoms in the amine and six in the acid segments. Other grades of nylon are 6.10, 6.12, 13.13, etc. The molecular structure of nylon '66' is shown in Fig. 8.9.

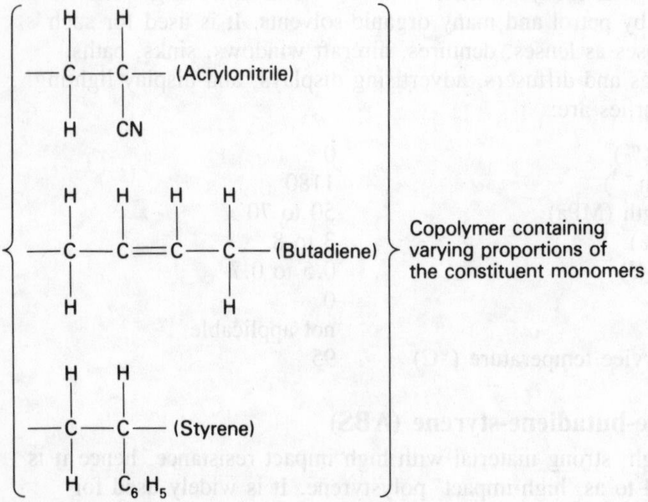

Fig. 8.8 Segment of the copolymer of ABS

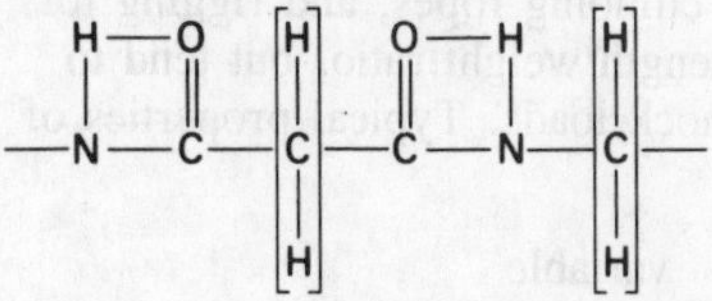

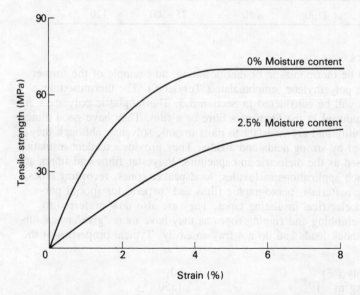

Fig. 8.9 Segment from nylon '66' polymer chain

This is just one segment from the molecular chain. Since the chain is linear, it gives rise to crystalline structures.

The nylon range of materials are strong, tough and flexible. They have good resistance to abrasion, but their dimensional stability, and electrical insulation properties are affected by the fact that nylon absorbs water, even from the atmosphere. This absorbed moisture acts as a plasticiser, lowering the stiffness, strength and hardness. The effect of moisture absorption on the strength of nylon '66' is shown in Fig. 8.10. The nylons show good resistance to most common solvents but, unfortunately, they do not weather well and degrade rapidly when used out of doors. Nylons can be moulded into such components as gears, valves, bearings, cams and surgical equipment. When used as bearings, the ability of nylon to absorb moisture from the atmosphere renders the need for a lubricant unnecessary. This is particularly useful in office machinery and food processing equipment. Nylons are also extruded and drawn into

Fig. 8.10 Effect of moisture content on the strength of nylon '66'

fibres for bristles, textiles, fishing lines, climbing ropes, and rigging for small boats. Nylon ropes have a high strength/weight ratio, but tend to fray easily and have little resilience to shock loads. Typical properties of nylon '66' are:

Crystallinity (%)	variable
Density (kg m^{-3})	1100
Tensile strength (MPa)	50 to 85
Elongation (%)	60 to 300
Impact value (J)	1.5 to 15.0
Tg (°C)	50
Tm (°C)	265
Maximum service temperature (°C)	120

Some further typical 'nylons' and their properties are listed in Table 8.2.

Table 8.2 Properties of typical nylons

Material	Properties			
	Density (kg m^{-3})	Tensile strength (MPa)	Elongation (%)	Maximum service temperature (°C)
Nylon 6	1100	70–90	60–300	120
Nylon 6.10	1100	60	80–230	120
Nylon 11	1100	50	75–300	120

Polyesters

These can be thermoplastic or thermosetting, an example of the former type being polyethylene teraphthalate ('Terylene'). The thermosetting polyesters will be considered in Section 8.5. Thermoplastic polyesters are usually produced in the form of a fibre or a film. They have good dimensional stability and are resistant to most organic solvents, although they are attacked by strong acids and alkalis. They provide excellent insulation and are used as the dielectric in capacitors. Polyester films and fibres are used in such applications as textiles, loudspeaker cones, recording tapes, draughting materials, photographic films and 'papers' for special purposes, and electrical insulating tapes. They are also now preferred to nylon for climbing and rigging ropes as they have more 'give' when subjected to shock loads and do not fray so easily. Typical properties of the fibre are:

Crystallinity (%)	60
Density (kg m^{-3})	1350
Tensile strength (MPa)	over 175

Elongation (%)	60 to 110
Impact value (J)	1.0
Tg (°C)	70
Tm (°C)	267
Maximum service temperature (°C)	69

Polyacetals

These materials are very strong and stiff with good creep resistance and fatigue endurance. They are beginning to replace metals in a number of applications because of their high crystallinity and clearly defined, high melting temperature. They are good electrical insulators and are resistant to alkalis and most common solvents. However they are attacked by mineral acids and degrade in strong sunlight, so that they are not recommended for outdoor use. Typical applications are: water pump impellers and housings for domestic appliances, extractor fan components, moulded instrument panels for road vehicles, plumbing fittings, bearings, cams, gears, hinges, window catches and door-lock components. Typical properties for *polyformaldehyde* are:

Crystallinity (%)	70 to 90
Density (kg m^{-3})	1410
Tensile strength (MPa)	50 to 70
Elongation (%)	15 to 75
Impact value (J)	0.5 to 2.0
Tg (°C)	−73
Tm (°C)	180
Maximum service temperature (°C)	105

Typical properties for *polyoxymethylene* are:

Crystallinity (%)	70 to 90
Density (kg m^{-3})	1410
Tensile strength (MPa)	60 to 70
Elongation (%)	15 to 70
Impact value (J)	0.5 to 2.0
Tg (°C)	−76
Tm (°C)	180
Maximum service temperature (°C)	120

Polycarbonates

These materials have good impact strength, good heat resistance, good dimensional stability, good electrical insulation properties, and good optical properties which are superior to perspex. Further, polycarbonates are extremely tough and have a much higher scratch resistance than any other transparent plastic. Polycarbonates are resistant to petroleum products and most solvents. Typical applications are: electrical insulators, capacitor dielectrics, lightweight, 'unbreakable' spectacle lenses, aircraft components and windows, and vehicle components. Typical properties are:

Crystallinity (%)	0
Density (kg m^{-3})	1200
Tensile strength (MPa)	60 to 70
Elongation (%)	60 to 100
Impact value (J)	10 to 20
Tg (°C)	150
Tm (°C)	not applicable
Maximum service temperature (°C)	130

Cellulose acetate

This is a tough, durable material which is not dangerously flammable as is the closely related cellulose nitrate (celluloid). Unfortunately, cellulose acetate is highly water absorbent and this adversely affects its dimensional stability and electrical insulation properties; it is unsuitable for use out of doors. The water absorption problem can be controlled to some extent by the addition of *butyrate*. All cellulosics are resistant to most household chemicals but are attacked by acids, alkalis, and alcohol. All cellulosics soften in boiling water. Cellulose acetate is used for lampshades and diffusers, machine guards, spectacle frames, screwdriver and other small tool handles, photographic (safety) film base, toys and many small components. It is easily moulded and takes a sharp and detailed impression. Typical properties are:

Crystallinity (%)	0
Density (kg m^{-3})	1280
Tensile strength (MPa)	24 to 65
Elongation (%)	5 to 55
Impact value (J)	0.7 to 7.0
Tg (°C)	120
Tm (°C)	not applicable
Maximum service temperature (°C)	70

8.5 Properties and applications of typical thermosets

The thermoplastics described in Section 8.4 soften every time they are heated above their glass transition temperatures and they can then be reshaped. Thermosetting plastics (thermosets) undergo chemical change (curing) during moulding and can never again be softened by heating (see Section 2.16). Thermosets are stronger, more rigid and more brittle than thermoplastics.

Phenol formaldehyde

Phenolic resins are never used by themselves but in conjuction with fillers and other additives which reduce the inherent brittleness, improve the mechanical and electrical properties, make them more amenable to moulding, and enhance their natural brown colour (see Section 8.1). The properties of phenolic mouldings vary widely depending upon the additives used. Typical applications are: electrical insulators, electrical plugs and sockets (industrial), handles, knobs, vehicle ignition system mouldings, and clutch and brake linings. Phenolic resins are also used in the manufacture of laminates such as 'Tufnol' (see Section 8.6). The dark colour of phenolic resins precludes their use for many domestic electrical applications. Typical properties of a wood flower filled phenolic moulding powder are:

Density (kg m^{-3})	1350
Tensile strength (MPa)	35 to 55
Elongation (%)	0 to 1
Impact value (J)	0.5 to 1.5
Maximum service temperature (°C)	75

Urea formaldehyde

The basic resin is hard, brittle, rigid and scratch resistant. Like phenol formaldehyde it is never used by itself but in conjunction with fillers and other additives (see Section 8.1). It is resistant to most solvents and household detergents. It has good electrical insulation properties and being virtually colourless, it can be coloured by the addition of pigments to suit any decorative requirements. For this reason it is widely used for domestic electrical equipment (plugs, sockets and switches) and most domestic utensils and toys. It is also used as a binder for foundry core-sands and for shell-moulding. Typical properties with a cellulose filler are:

Density (kg m^{-3})	1500
Tensile strength (MPa)	50 to 75
Elongation (%)	0 to 1
Impact value (J)	0.3 to 0.5
Maximum service temperature (°C)	75

Melamine formaldehyde

This material is similar to urea formaldehyde, but is more resistant to heat and is less water absorbent. This not only improves its electrical properties but makes it suitable for tableware. Typical properties with a cellulose filler are:

Density (kg m^{-3})	1500
Tensile strength (MPa)	56 to 80
Elongation (%)	0 to 0.7
Impact value (J)	0.2 to 0.5
Maximum service temperature (°C)	100

Epoxides

Epoxy resins are used for bonding glass fibre fillers as described in Section 8.7. They are resistant to water and most reagents and have excellent electrical insulation properties. Therefore epoxides are also widely used as a casting material for small components and as a 'potting' material for sealing electrical equipment such as transformers and chokes. Their use as the basis of high strength adhesives will be considered in *Engineering Materials: volume 2*. Typical properties of an unfilled resin casting are:

Density (kg m^{-3})	1150
Tensile strength (MPa)	35 to 50
Elongation (%)	5 to 10
Impact value (J)	0.5 to 1.5
Maximum service temperature (°C)	200

Polyesters (unsaturated)

The mechanical properties of unsaturated polyesters vary quite considerably according to the polymer used. These materials have good impact resistance and surface hardness. They are also resistant to water and most common solvents but are attacked by acids and alkalis. Their low water absorption and resistance to weathering make them an excellent binder for use with glass fibre reinforcement (see Section 8.7) for mouldings ranging from domestic baths to boat hulls for pleasure and small commercial and naval craft. Typical properties of the unfilled (not reinforced) resin are:

Density (kg m^{-3})	1120
Tensile strength (MPa)	55
Elongation (%)	2
Impact value (J)	0.5 to 1.0
Maximum service temperature (°C)	200

Polyesters (alkyds)

These materials have good heat resistance and excellent electrical insulation properties. They have good dimensional stability and are unaffected by water and most organic solvents. This makes them suitable for mouldings for high-voltage insulators in television sets, and for mouldings for the electrical equipment of road vehicles and aircraft. Alkyd resins are also used as the basis for the paint systems used on cars and domestic appliances. Typical properties are:

Density (kg m^{-3})	2000
Tensile strength (MPa)	25
Elongation (%)	Nil
Impact value (J)	0.25 to 0.5
Maximum service temperature (°C)	150

Comparing the properties for the thermosets with those for the thermoplastics shows that the main advantages of the former over the latter lie in their superior abrasion resistance (hardness), rigidity, and high maximum service temperature. However their impact value is lower and they tend to be brittle and break more easily. Because curing takes place in the mould, the moulding time cycle takes longer and processing is more costly than for thermoplastic materials.

8.6 Laminated plastic 'Tufnol'

Fibrous material such as paper, woven cotton and woollen cloth, woven glass fibre cloth, or woven asbestos cloth (not now recommended as it is a potential health hazard) can be used to reinforce phenolic and epoxy resins. Sheets of these fibrous reinforcement materials are impregnated with the resin and they are then laid up between highly polished metal plates in hydraulic presses. The thickness of the finished sheet is determined by the number of layers of impregnated reinforcement in the laminate. Each layer of reinforcement is rotated through ninety degrees of arc so as to ensure uniformity of mechanical properties. The laminates are then heated under pressure until they become solid sheets, rods or tubes.

Laminated composites can be machined dry with ordinary engineers' tools using a low value of rake angle and fairly high cutting speeds. However, the material is rather abrasive and carbide tooling can be used to advantage, particularly when machining this material on a production basis. Although not toxic, the dust and fumes generated during machining is unpleasant and adequate extraction should be provided. Care must be taken in the design and use of plastic laminates because of the 'grain' of the material which makes it behave in a similar manner to plywood.

'Tufnol' composites are widely used for making bearings, gears and other engineering products which have to operate in hostile environments or where adequate lubrication is often not possible as in food processing and office machinery. It is also used for making insulators for heavy duty electrical equipment.

8.7 Glass fibre reinforced plastics

Glass fibre reinforced plastic mouldings are used in the manufacture of many articles of complex shape requiring a high strength/weight ratio and resistance to environmental attack. Typical examples range from fishing rods to naval vessels, private cruisers and yachts. Safety helmets, domestic baths, casting patterns, copy milling models, machine guards, light aircraft and gliders are all made from glass fibre reinforced plastics. Smaller products are hot moulded in semi-automatic processes, whilst larger mouldings are laid up individually by skilled craftspersons. Polyester resins are the most widely used for bonding the fibres together, but epoxy resins are used where maximum strength is required and the higher cost can be justified.

The glass used for fibre manufacture is not the same as that used for window glass. The composition of the glass depends upon the application for which the moulding is being produced.

E-glass (electrical grade) is used for the manufacture of high grade printed circuit boards. It has excellent electrical insulation properties and dimensional stability. As well as for electrical mouldings it is also widely used for general purpose mouldings.

C-glass is used for chemical plant mouldings. It is low in aluminium oxide and calcium oxide content, and has good resistance to acid attack. It is also used for fibre glass surfacing mats which must resist environmental attack and which must protect the substrata of the moulding.

S-glass is a high strength fibre produced in continuous filaments for weaving into mats and fabrics for the manufacture of pressure vessels and boat hulls.

M-glass has an exceptionally high tensile modulus, but because of its high cost it has a restricted range of applications.

Fibre glass has entirely different mechanical properties from bulk glass as used for windows. Because of their very high surface area, glass fibres can have a tensile strength considerably greater than bulk glass in addition to being of different composition. The properties of E-glass and S-glass compared with high carbon steel wire (piano wire) are listed in Table 8.3. Glass fibres are usually drawn through platinum dies having 204 holes. The 204 filaments so formed are gathered together into one *basic strand*. These strands are then gathered together into *rovings* of 8 to 408 basic strands, or they can be twisted and plied into yarns. Fibre diameters can range from 0.0025 mm to 0.025 mm.

As well as the direct influence of the fibre content on the tensile

Table 8.3 Properties of typical glass fibres

Material	Relative density	Tensile strength (GPa)	Tensile modulus (GPa)	Specific strength (GPa)(1)	Specific modulus (GPa)(2)
E-Glass	2.55	3.5	74	1.4	29
S-Glass	2.50	4.5	88	1.8	35
Steel wire (for comparison)	7.74	4.2	200	0.54	26

1. Specific strength = $\dfrac{tensile\ strength}{relative\ density}$

2. Specific modulus = $\dfrac{tensile\ modulus}{relative\ density}$

3. Typical composition for E-glass

Silicon dioxide	52−56%
Calcium oxide	16−25%
Aluminium oxide	12−16%
Boron oxide	8−13%
Sodium and potassium oxides	1%
Magnesium oxide	0−6%

modulus (flexual stiffness) and the tensile strength of a GRP composite, the strength of the composite is also influenced by the orientation of the fibres.

Parallel yarns. All the glass strands are laid parallel to each other to provide unidirectional reinforcement for such applications as fishing rods.

Woven cloth. Half the strands are laid at right angles to the other half and locked together by weaving. This provides bi-directional reinforcement and is suitable for boat hulls where high strength is required over large areas of gently changing curvature.

Chopped strand mat. Short strands of glass fibre are arranged in a totally random manner to form an isotropic reinforcement, that is, the reinforcement is equal in all directions. Chopped strand mat is used where strength has to be combined with sharp curves and complex shapes as in safety helmets and machine guards.

The amount of reinforcement which can be used depends upon the orientation of the reinforcement. With long strands laid up parallel to each other the *reinforcement area fraction* can be as high as 0.9. The reinforcement area fraction is the cross-sectional area of reinforcement divided by the total cross-sectional area as shown in Fig. 8.11. With

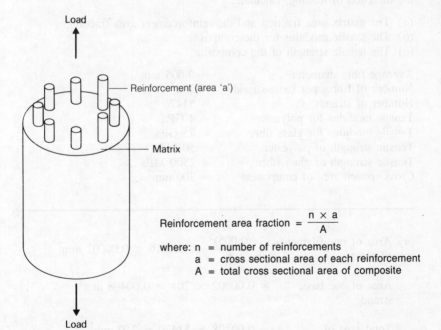

Reinforcement area fraction = $\dfrac{n \times a}{A}$

where: n = number of reinforcements
a = cross sectional area of each reinforcement
A = total cross sectional area of composite

Fig. 8.11 Reinforced composite

Table 8.4 Properties of GRP composites

Material	Reinforcement (weight %)	Tensile strength (MPa)	Tensile modulus (GPa)
Chopped strand mat	10–45	45–180	15–15
Plain weave cloth	45–65	250–375	10–20
Long fibres (uniaxially loaded)	55–80	500–1200	25–50

woven strand fabrics the reinforcement area fraction can be as high as 0.75. With chopped strand mat the reinforcement area fraction is substantially reduced but a figure of 0.5 should be considered the minimum for satisfactory reinforcement. Obviously the lower the reinforcement area fraction, the weaker the composite produced. Table 8.4 shows the effect of the type of reinforcement on the strength of the composite, whilst Example 8.1 shows how the 'reinforcement area fraction', tensile modulus and tensile strength of a composite can be calculated.

Example 8.1 Given the following data for a glass-reinforced polyester component of rectangular cross-section in which the strands lie parallel to the direction of loading, calculate:

(a) The matrix area fraction and the reinforcement area fraction.
(b) The tensile modulus for the composite.
(c) The tensile strength of the composite.

Average fibre diameter	= 0.005 mm
Number of fibres per basic strand	= 204
Number of strands	= 51470
Tensile modulus for polyester	= 4 GPa
Tensile modulus for glass fibre	= 75 GPa
Tensile strength of polyester	= 50 MPa
Tensile strength of glass fibre	= 1500 MPa
Cross section area of component	= 300 mm^2.

(a) Area of one filament of glass $= \dfrac{(0.005)^2}{4} \times 3.1416 = 0.00002$ mm^2

Area of one basic strand $= 0.00002 \times 204 = 0.00408$ mm^2

Total area of reinforcement $= 0.00408 \times 51470 = 210$ mm^2

$\therefore$ The reinforcement area fraction $= \dfrac{\text{total area of reinforcement}}{\text{total component area}}$

$$= \frac{210}{300}$$

$$= \underline{\underline{0.7}}$$

But the matrix area $=$ total area $-$ reinforcement area
$= 300 - 210$
$= 90 \text{ mm}^2$

$\therefore$ The matrix area fraction $= \dfrac{\text{total matrix area}}{\text{total component area}}$

$$= \frac{90}{300}$$

$$= 0.3$$

(b) Modulus of composite $=$ (modulus of matrix $\times$ matrix area fraction)
   $+$ (modulus of reinforcement $\times$ reinforcement area fraction).
   $= (4 \times 0.3) + (75 \times 0.7)$
   $= 1.2 + 52.5$
   $= \underline{\underline{53.7 \text{ GPa}}}$

(c) Tensile strength of composite $=$ (Tensile strength of matrix $\times$ matrix AF)
   $+$ (tensile strength of reinforcement $\times$ reinforcement AF)
   $= (50 \times 0.3) + (1500 \times 0.7)$
   $= 15 + 1050$
   $= \underline{\underline{1065 \text{ MPa}}}$

The above example indicates the importance of the reinforcement in any composite material when the contribution of the reinforcement is compared with the contribution of the matrix. If the calculation is repeated for a range of reinforcement area fractions, the curves shown in Fig. 8.12 will be obtained. Fig. 8.12(a) plots the tensile modulus against reinforcement area fraction, and Fig. 8.12(b) plots the tensile strength of the composite against the reinforcement area fraction.

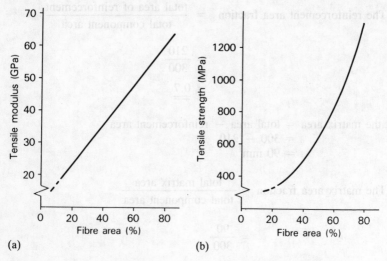

Fig. 8.12 Effect of reinforcement area on a typical GRP composite

9 Bearing materials

9.1 Material surfaces

A perfect plain surface is a geometrical plane devoid of irregularities. Such a surface never exists in practice. In reality, all material surfaces — no matter how smooth they may appear to the unaided eye — will have a texture similar to that shown in Fig. 9.1. *Real surfaces*, as defined in BS 1134, are the actual physical surfaces separating the component from surrounding space. Such surfaces have three main characteristics:

(a) *Waviness.* That component of surface texture upon which roughness is superimposed. Waviness may be caused by such factors as machine or work deflections, vibrations, chatter, heat-treatment, or warping strains.

(b) *Roughness.* Irregularities of the surface texture which are inherent in the production process (tooling marks) but excluding waviness and errors of form.

(c) *Lay.* The direction of the predominant surface pattern, ordinarily determined by the production method used.

All these characteristics will have a profound effect upon the performance of any two real surfaces which are in contact with each other to form a bearing. The life of these mating surfaces, when used for such applications as shafts and bearings, is dependent upon the surface texture as well as upon the physical characteristics of the materials from which the mating components are made and any lubricant used.

A rough surface with large peaks and valleys will have less contact

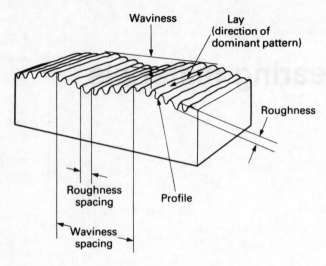

Fig. 9.1 Surface characteristics and terminology

area and will wear more quickly than a smooth surface. Unfortunately, the smoother and more geometrically accurate a surface becomes, the more costly it is to produce. Further, it is no use specifying a dimensional tolerance to a process whose inherent surface roughness lies outside the limits of that tolerance as shown in Fig. 9.2(*a* and *b*). Even two surfaces having the same roughness index can have different wearing characteristics, as shown in Fig. 9.3(*a* and *b*). They both have the same height of peaks and the same depth of valleys with the same spacing, but it is obvious that the first surface will wear less quickly than the second surface under the same conditions of service.

9.2 Wear

Figure 9.4 shows a magnified section through two mating surfaces from which any form of lubricant has been removed. The entire load on the surfaces is supported only on the high spots of the peaks. These high spots are called *asperities*. Since the contact surface area of the asperities may be very small, the contact pressure, even under light loads, will be very great and *plastic deformation* will occur in ductile materials. This will result in the contact area increasing, and the contact pressure decreasing, until the material can support the applied load.

Abrasive wear

Figure 9.4 shows how the asperities become interlocked under load. Before the surfaces can slide over each other, the asperities have to shear along the planes AA or BB or both. The resisting (friction) force is the

207

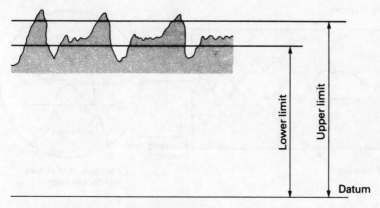

(a) Limits of size and process mismatched

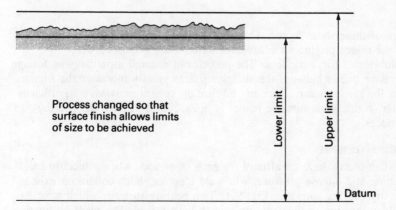

Process changed so that
surface finish allows limits
of size to be achieved

(b) Process suitable for limits of size

Fig. 9.2 Effect of surface finish on production process

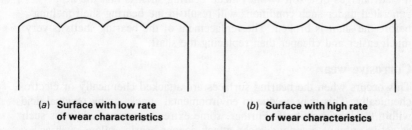

**(a) Surface with low rate
of wear characteristics**

**(b) Surface with high rate
of wear characteristics**

Fig. 9.3 Wear characteristics

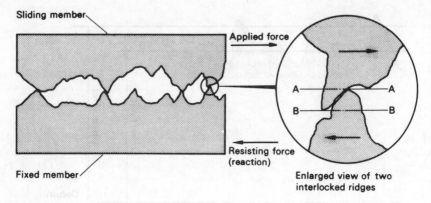

Fig. 9.4 Friction and wear

sum of these shear forces. The continual interlocking and shearing away of the ridges produces the *abrasive wear* which occurs between unlubricated (dry) surfaces. The particles of sheared asperities and foreign abrasive matter between the sliding surfaces greatly increases the friction and the rate of wear. Thus any lubrication system must have an efficient filter so that such abrasive matter is not recycled through the bearing surfaces.

Adhesive wear

Certain metals have an affinity for each other and, when subject to local heating and intense pressure, will weld together. Such conditions exist in the dry bearing shown in Fig. 9.4. Two bearing surfaces which have welded together in this way are said to have *seized*. The resisting forces created and the wear which takes place under such conditions, as the welded junctions shear, are very much greater than when the shearing forces are limited to the interlocking asperities. This is *adhesive wear*.

In extreme cases the shaft may break in torsion or the bearing may be ripped from its housing before shear of the junctions takes place. One of the advantages of a soft (white) metal bearing shell is that the heat generated under such conditions will result in the bearing shell melting before the shaft is broken. The replacement of the bearing shells is very much easier and cheaper than replacing the shaft.

Corrosive wear

This occurs when the bearing surfaces are attacked chemically or electro-chemically. This may be due to environmental attack, or by residual acids within the lubricant itself. Further, some extreme pressure additives such as active sulphur compounds can attack copper bearing alloys such as phosphor bronze.

Surface fatigue

The bearing surfaces considered so far have been *sliding*. Figure 9.5 shows what happens in *rolling* bearings. It can be seen from Fig. 9.5(*a*) that when a perfect cylinder is supported by a plain surface only 'line contact' exists. Since a line has no area, any load on the roller results in infinite contact pressure. In practice the roller, or the supporting surface, or both, will collapse slightly due to *elastic deformation* until the contact area is increased (Fig. 9.5(*b*)), and the contact pressure correspondingly decreased sufficiently for the bearing materials to sustain the load.

The constant flexing of the rolling surfaces due to elastic deformation can result in cyclical stresses which may exceed the fatigue resistance of the material. This can result in fatigue cracks below the surface of the material at a depth of 0.25 mm. These initial cracks usually propagate parallel to the surface, causing flaking (spalling) of the material from the surface. High surface hardness, and a highly polished surface free from scratches and fissures (hardening cracks and machining marks) will reduce the rate of fatigue wear.

9.3 Sliding friction

Friction is defined as *the resistance which opposes the motion of one surface as it moves across another*. It has already been shown that this resistance to movement is caused by the interlocking and welding of the asperities of the mating surfaces. Consider Fig. 9.6. The true (real) area of contact (A_r) between the two surfaces is much smaller than the apparent (theoretical) contact area and can be approximated from the following equation:

$$A_r = \frac{F_n}{Y_p}$$

where: A_r = real contact area
   F_n = force normal to the contact area (load)
   Y_p = yield pressure causing plastic deformation of the contact areas.

The force (F_s) tending to cause relative sliding between the two surfaces is the force required to shear all the junctions and will be proportional to the shear strength of the material at the junctions and proportional to the total area in shear (which is also the real area of contact). Thus:

$$F_s = A_r \times S_u$$

where: F_s = shear force to cause movement
   A_r = total (real) area in shear
   S_u = ultimate shear strength of the material at the junctions.

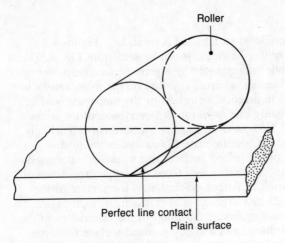

Perfect line contact
Plain surface

(a) Theoretical rolling contact.

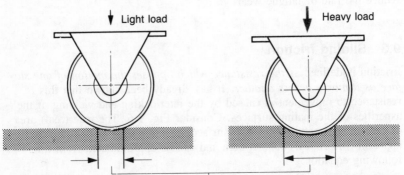

Light load · Heavy load

Hard, rigid, metal wheels bite into the floor until the area of contact will support the load.

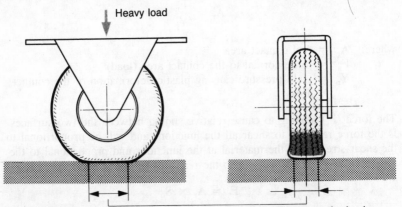

Heavy load

Rubber tyres distort under load until the area of contact will support the load.
(b) Practical rolling contact
Fig. 9.5 Elastic deformation and flexure of rolling bearings

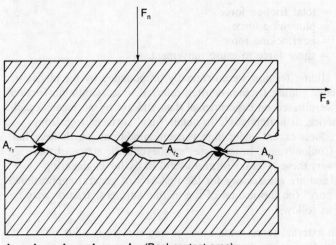

$A_r = A_{r_1} + A_{r_2} + A_{r_3} \ldots A_{rn}$ (Real contact area).

F_n = normal force

F_s = shear force to cause movement.

Fig. 9.6 Real contact during sliding

Dividing F_s by F_n:

$$\frac{F_s}{F_n} = \frac{A_r \times S_u}{A_r \times Y_p} = \frac{S_u}{Y_p}$$

But from Fig. 9.7:

$$\frac{F_s}{F_n} = \mu$$

where: μ = the coefficient of limiting friction.

Therefore:

$$\mu = \frac{S_u}{Y_p}$$

That is, the coefficient of limiting friction μ is the total shear strength of the junctions divided by the yield pressure (hardness) of the softer material. In practice, the sliding friction may involve additional effects such as the ploughing of the softer material by the asperities of the harder material, and the interlocking of the surface irregularities. Therefore the total friction force will be equal to these three components.

$$F_t = F_s + F_p + F_i$$

where: F_t = total friction force
        F_p = ploughing force
        F_i = interlocking force
        F_s = shear force to cause movement.

The laws of sliding friction were originally propounded by the physicist, Coulomb, hence sliding friction is often referred to as 'Coulomb friction'. These laws were based on experimental data and are by no means exact since, in practice, it is impossible to obtain 'dry' surfaces free from oxidation products except under high vacuum conditions which were not available to Coulomb and, in any case, are beyond the scope of this book. However, these laws do form a useful basis for engineering calculations, and are satisfactory for the majority of general applications. The laws of 'dry' (no lubrication of the mating surfaces) friction may be summarised as follows:

(a) When an external force tends to cause one surface to slide over another surface, a reactionary frictional force is set up, acting tangentially to the surfaces so as to oppose the motion.

(b) There is a limiting value to the force of friction beyond which it cannot rise. If the externally applied force exceeds this value sliding will commence.

(c) The force required to start sliding is greater than that to maintain it. Static friction is greater than kinetic friction.

(d) The limiting value of the frictional force maintains a constant ratio to the normal reaction between the surfaces. This ratio is called the coefficient of limiting friction, and is denoted by the greek letter μ (mu).

(e) The coefficient of limiting friction depends upon the mating surfaces, their surface texture, surface contamination, and the physical properties of the materials.

Figure 9.7 gives some typical values for μ for various combinations of materials and provides a basis for Example 9.1.

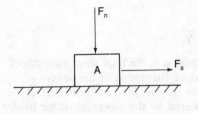

Materials (Dry)	μ
Cast iron on brass	0.15
Steel on brass	0.15
Steel on phosphor bronze	0.12
Steel on cast iron	0.20
Ferodo brake lining on cast iron (for comparison)	0.60

Fig. 9.7 Sliding friction

Example 9.1 Calculate the force to just move the block 'A' if the normal force is 32N and the coefficient of friction μ is 0.125.

$$\frac{F_s}{F_n} = \mu \quad \text{where: } F_s = \text{force to move 'A'}$$
$$F_n = 32 \text{ N}$$
$$\mu = 0.125$$

$$F_s = \mu F_n$$
$$= 0.125 \times 32$$
$$= 4.0 \text{ N}$$

9.4 Rolling friction

Rolling friction occurs when a cylinder or sphere rolls over a plain surface. The load is assumed to act through the centre of the roller or sphere. Lubrication is necessary because, as has been previously stated, some elastic deformation occurs at the point of contact so that true rolling does not actually occur and a small amount of slip is present. However this is very slight and rolling friction is very much less than sliding friction. Rolling friction is proportional to the applied load and inversely proportional to the diameter of the roller or sphere (see Fig. 9.8). The coefficient for rolling friction is the same for both static and kinetic friction, whereas for sliding friction, the coefficient for static friction is very much greater than the coefficient for kinetic friction. In rolling bearings most of the friction losses and heating effects are due to the elastic hysteresis losses within the component materials.

$f_r = F_r/F_n$
where:
f_r = coefficient of rolling friction
F_r = rolling friction force
F_n = Load

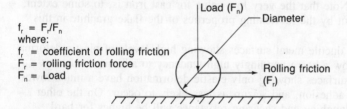

Fig. 9.8 Rolling friction

9.5 Surface contamination

When metals are exposed to air they are always covered by a film of metal oxide, together with a film of moisture and, frequently, with surface layers of other chemical compounds and absorbed gases. These greatly affect the adhesion and friction between the moving surfaces and, as long as these films are present, the coefficient of limiting friction rarely exceeds 1.5. However for perfectly clean metal surfaces prepared under high vacuum conditions the coefficient of friction may reach very high values. Under such conditions adhesion will be very strong and a metallic bond may form between the surface asperities resulting in cold-welding and seizure. The coefficient of friction for diamond under normal atmospheric conditions is 0.05, but when it is outgassed in a vacuum its coefficient of friction rises tenfold to 0.5. Another example is graphite. This material has a low coefficient of friction under normal atmospheric conditions not only because of its laminar structure but also because of its surface film of water and dissolved gases. Under such conditions the coefficient of friction for a graphite surface is 0.1, but when dried and outgassed in vacuum its coefficient of friction rises to between 0.5 and 0.8 and it rapidly wears away to a fine dust.

9.6 Similar and dissimilar materials

Amontons laws of friction state that the coefficient of friction is independent of load and that the *real contact area* increases directly with an increasing load. This increase occurs during sliding between the surfaces because the area of the junctions increases as the result of the combined effect of plastic deformation and shearing. This applies particularly to pure metals. Further, the interface between two surfaces of the same or similar metals shows a greater coefficient of friction and a greater adhesion than an interface between two dissimilar metals. Again, homogeneous materials tend to show greater adhesion than heterogeneous materials. For example, cast iron on steel has a coefficient of friction of only 0.3, whereas nickel, pure iron (ferrite) and austenitic stainless steel (all highly homogeneous) show coefficients of friction within the range 1.2 to 1.5. Note that the very low value for cast iron is, to some extent, brought about by the lubricating properties of the flake graphite in this material.

Soft and ductile metal surfaces showing high levels of plastic deformation also show a correspondingly high tendency to adhesion, whereas hard metal surfaces showing only elastic deformation have a much lower tendency to adhesion, and seizure is less likely to occur. On the other hand, the ploughing and shearing resistance will be higher for hard materials.

Polymer bearing materials behave very much like metals. Deformation occurs at the junctions and strong adhesion may take place. However, unlike metals, this is viscoelastic deformation and depends upon the surface texture, the load, and the duration of loading. Thus there is a lower level of junction growth and this accounts for the fact that, unlike metals, the friction of polymer materials does not increase linearly with the load and that the coefficient of friction actually decreases with increasing load.

Adhesion tends to be strong between polymer materials and their coefficients of friction tend to be fairly high, ranging from 0.3 to 0.5. The exception is *polytetrafluoroethylene* which has the lowest coefficient of friction of any known substance. A typical value under normal atmospheric conditions is 0.04.

9.7 Heating effects

When unlubricated surfaces slide over each other the mechanical energy required to overcome frictional resistance is converted into heat energy. This is not uniformly distributed over the apparent contact area but is concentrated mainly at the asperities where it is generated. This may raise their temperatures above their melting points and accounts for the formation of welded junctions. The surface temperature depends upon a number of factors such as the relative load and speed, thermal conductivity of the surface materials, the mass of metal available to conduct the heat away, and the coefficient of friction. Polymer materials present a particular problem as they are very good heat insulators and thermal conductivity through the bearing is low. Further, they usually have low softening and melting temperatures.

Frictional heat can produce very smooth surface finishes if the mating surfaces are properly 'run in'. Surface material is transferred from the peaks to the valleys, and 'hot-spots' result in softened or molten metal being spread over the surface where it cools quickly to form a characteristic layer of a glassy, amorphous appearance called the *Beilby layer*.

9.8 Lubrication

The friction between two surfaces has already been shown to result from the interlocking, shearing and welding of asperities of the surfaces. If the surfaces can be separated by a lubricant, then the friction and wear becomes negligible. For example it is far harder to drag a small boat over the shingle of the beach than to move it once it is floating and there is a layer of water acting as a lubricant between the boat and the shingle. Further, the wear on the bottom of the boat will be negligible once it is in the water compared with the damage that will be done dragging it over the dry shingle beach.

There are two mechanisms of lubrication, *fluid lubrication* (as described above), and *boundary layer lubrication*. Full fluid lubrication occurs when the bearing surfaces are completely separated by a thick film of lubricant capable of supporting the loads applied to the bearing. Under these conditions there is no direct contact between the material surfaces, as shown in Fig. 9.9(*a*), and the coefficient of friction becomes very low (0.001 to 0.01) being independent of the bearing materials and is determined only by the viscosity of the lubricant.

Boundary layer lubrication occurs when the layer of lubricant adhering to the bearing surfaces is only a few molecules thick as shown in Fig.

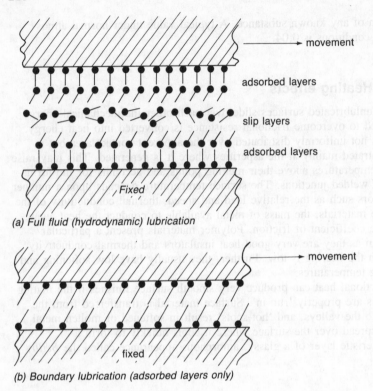

(a) Full fluid (hydrodynamic) lubrication

(b) Boundary lubrication (adsorbed layers only)

Fig. 9.9 Lubrication

9.9(*b*). This is often only comparable with the valleys and asperities of the bearing surfaces. Lubrication will be minimal and there will be some contact between the asperities resulting in high rates of wear. This occurs during the starting and stopping of machines and engines when oil flow is negligible or is non-existent and only a residual film is present.

Between these two extremes there is the condition when oil is commencing to flow into the bearing and the bearing surfaces are partly in contact and partly separated by the lubricant. This is referred to as *partial fluid lubrication*. Figure 9.10 shows how the coefficient of friction can change with these different conditions.

The viscosity of the lubricant changes with temperature and a compromise has to be achieved between an oil which is 'thin' enough at low (starting up) temperatures to avoid excessive drag, yet does not become so thin under the higher temperatures of operating and overload conditions that it cannot protect the bearing. This has resulted in the development of 'multi-grade' oils.

Mechanisms are usually operated under full fluid lubrication conditions since the friction is very low and wear is minimal. Boundary

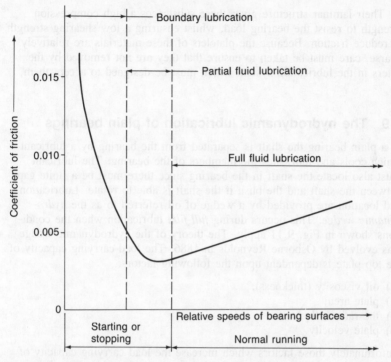

Fig. 9.10 Coefficient of friction changes with service conditions

lubrication properties are provided by adding fatty acids such as carboxyl to a lubricating oil which has been refined from a petroleum crude. The reaction of the fatty polar groups with the metal surface results in a monomolecular layer which adheres strongly to the surface with a long hydrocarbon chain oriented outward, perpendicular to the surface as previously shown in Fig. 9.9. Under high pressures and speeds there is an excessive build up of heat and the boundary film melts and breaks down. Special additives which will form films such as sulphides, chlorides and phosphides on the metal surface are used. Such films have sufficient shearing strength and high temperature resistance to withstand severe service conditions. These are called *extreme pressure additives* and lubricants which contain them are called *extreme pressure lubricants.* They are ineffective on inert metal surfaces such as chromium, but are widely used where sliding and rolling occur at the same time under heavy load conditions as in the hypoid bevel gears of road vehicle final drives. As fast as the boundary film is rubbed away, it is replaced by the chemical reaction between the additive and the bearing surfaces. Improved boundary lubrication and reduced wear may also be provided by the addition of inorganic materials such as graphite or molydenum disulphide to the lubricant.

Their laminar structure gives these substances a high compression strength to resist the bearing load, whilst ensuring a low shearing strength to reduce friction. Because the platelets of these materials are relatively coarse, care must be taken to ensure that they are not removed by the filters in the lubrication system which must be designed to accept them.

9.9 The hydrodynamic lubrication of plain bearings

In a plain bearing the shaft is separated from the bearing by a lubricant which cools and lubricates the members of the bearing. The lubricant must also locate the shaft in the bearing since there must be a finite gap between the shaft and the bush if the shaft is able to rotate. Lubrication and location are provided by a wedge of oil referred to as the *hydro-dynamic wedge*. This occurs during *full-film* lubrication when the conditions shown in Fig. 9.11 apply. The theory of the hydrodynamic wedge was evolved by Osborne Reynolds in 1886. The load-carrying capacity of the top plate is dependent upon the following factors:

(*a*) oil viscosity (thickness);
(*b*) plate area;
(*c*) the ratio h_1/h_2;
(*d*) plate velocity.

Unfortunately those factors which increase the load carrying capacity of the bearing also increase the fluid friction (drag). However Reynolds evolved two important mathematical equations to support his theory and these showed that whilst the one for friction was a first order (linear) equation, the one for load carrying capacity was a second order (quadratic) equation. It is therefore possible to 'play off' these two equations against each other and obtain an optimum set of conditions which

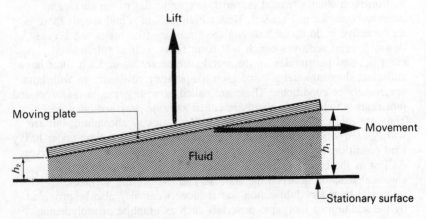

Fig. 9.11 The hydrodynamic wedge

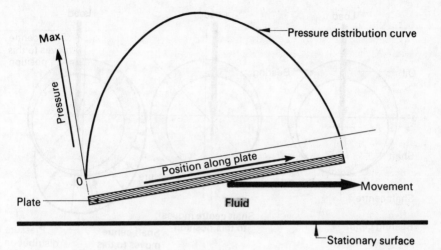

Fig. 9.12 Plain bearing-pressure distribution

will give maximum lift for minimum drag. The pressure distribution curve is shown in Fig. 9.12, and it can be seen that maximum lift does not occur at the point of minimum gap, as is often assumed, but at a point along the plate.

In practice it is the plain journal bearing which makes use of the converging films theory of Reynolds. Figure 9.13(a) shows the position of the shaft centre in a plain cylindrical bearing under 'rest', 'start' and 'run' conditions. It can be seen that the shaft centre moves about due to the changing and uneven distribution of lubricant pressure. Therefore this simple bearing is unsuitable for applications where the axis of the shaft must remain constant as in a machine tool spindle. Further, the introduction of oil grooves upsets the pressure distribution still further, but without them uniform lubrication of the bearing is not possible. For precision bearing applications constant centre-line bearings have been developed. The principle of such a bearing is shown in Fig. 9.13(b). The pressure system created by the rotating shaft causes the tilting pads to 'float' until the pressure lobes are uniformly distributed. Any disturbing force on the shaft upsets the balance of the system causing local pressure increases which oppose the disturbance. Thus the shaft is kept centred in the bearing.

9.10 Hydrostatic bearings

In this type of bearing the shell is perforated with fine holes through which air or oil is forced under high pressure. There is an appreciable gap between the shaft and the bearing shell and the shaft floats on a

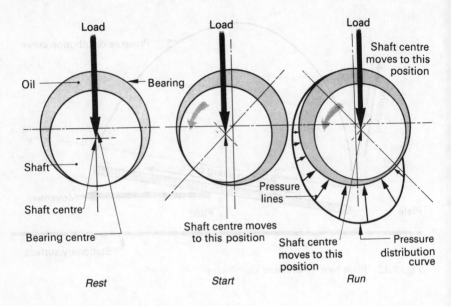

(a) Simple journal bearing

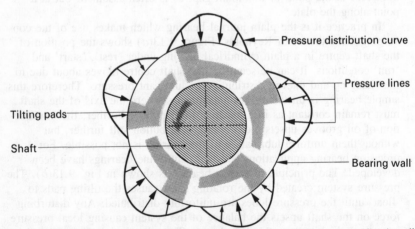

The pressure system created by the rotating shaft causes the tilting pads to 'float' until the pressure lobes are uniformly distributed

Any disturbing force on the shaft upsets the balance of the system causing local pressure increases that oppose the disturbance. Thus the shaft is kept centred in the bearing

(b) Constant centre-line bearing

Fig. 9.13 Plain journal bearing

221

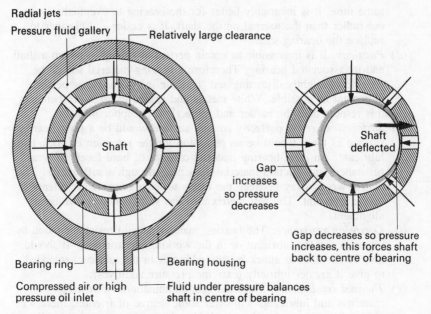

Fig. 9.14 The hydrostatic bearing

cushion of air or oil in a similar manner to a hovercraft. The principle of the bearing is shown in Fig. 9.14.

9.11 Requirements of bearing materials (sliding)

There are essentially two types of bearings. Those where the moving parts slide over each other as in plain bearings and machine-tool slideways, and those where the moving parts roll over each other as in ball and roller bearings.

Sliding surface bearings work under full fluid film lubrication conditions which separate the moving surfaces. However some direct surface to surface contact occurs from time to time when the lubrication film breaks down under overload conditions or when sliding starts or stops. Materials for sliding bearings require the following properties.

(a) *Coefficient of friction* (μ). This should be kept as low as possible to avoid wear, wasting energy and generating excessive heat in the bearing.
(b) *Strength*. The bearing must have sufficient compression strength to support the shaft and any load that may be applied to the shaft in service.
(c) *Wear resistance*. The bearing material must resist wear but, at the

same time, it is invariably better for the bearing to eventually wear out rather than the journal on the shaft. It is easier and cheaper to replace the bearing shell.

(d) *Plasticity.* It is impossible to obtain perfect alignment between a shaft and its associated bearing. Therefore a bearing material should be capable of slightly distorting and 'bedding in' to ensure as perfect alignment as possible. White metals and leaded bronzes are better in this respect than the harder and more rigid phosphor bronzes.

(e) *Surface texture.* A perfectly smooth surface would be a poor bearing surface as there would be no provision for the retention of pockets of lubricant. An ideal bearing material consists of hard facets of wear resistance, anti-friction material dispersed through a soft matrix. The matrix wears away between the facets to leave pockets for retention of the lubricant. The soft matrix also has sufficent 'give' to assist alignment.

(f) *Corrosion resistance.* The bearing material should resist corrosion by impurities in the lubricant or in the working environment. It should also be resistant to attack by any additives in the lubricant intended to give it greater lubricity (extreme pressure additives).

(g) *Thermal conductivity.* Since the best combinations of bearing materials and lubricants will offer some degree of friction, there will always be some energy loss in the bearing and some corresponding rise in temperature when the shaft is rotating. To keep the temperature rise to a minimum it is necessary to dissipate the heat energy as quickly as possible. This heat energy can only be dissipated through the lubricant and by conduction through the walls of the bearing. If the heat energy generated in the bearing is not conducted away quickly enough the temperature rise could reach the melting point of the bearing material with disastrous results.

9.12 Requirements of bearing materials (rolling)

The requirements for materials suitable for rolling bearings are somewhat different from those just described for sliding bearings. To appreciate the requirements of the materials used in rolling bearings it is necessary to consider the construction and use of such bearings. Rolling, or 'anti-friction', bearings consist of hardened balls or rollers arranged to roll between hardened steel rings called 'raceways'.

Ball bearings

Such bearings consist of four main components:

 (i) the inner race which is a light press fit on the shaft;
 (ii) the precision ground balls;
 (iii) the cage which keeps the balls equally spaced;
 (iv) the outer cage which is a light press fit in the bearing housing.

Theoretically, the balls should roll on the races without any slip taking place. Since sliding should not take place lubrication should not be necessary. However, because of the elastic deformation of the bearing elements under load, this is not the case and true rolling does not occur, therefore lubrication is essential. It is also necessary to lubricate the balls in their cage where sliding friction occurs.

Ball bearings are normally selected for high speed applications that are only lightly loaded (theoretically they only offer point support). They are less critical in their installation than roller bearings as they tend to be self aligning. Figure 9.15(a) shows some typical ball bearings and Fig. 9.15(b) shows a typical application.

Roller bearings

These are built up similarly to ball bearings with the exception of needle roller bearings (which are used where room is limited) where the rollers bear directly onto the shaft journal which is hardened and ground and there is no inner race. A typical needle roller bearing is shown in Fig. 9.16. Since roller bearings offer line support they are capable of handling heavier loads than ball bearings. They are widely used in supporting machine tool spindles subject to heavy cutting forces. In this connection tapered roller bearings are used in opposed pairs, to provide axial as well as radial restraint. Figure 9.17(a) shows the geometry for a tapered roller bearing where all axes and surfaces meet at a common point when they are projected. Figure 9.17(b) shows a typical application of roller bearings to a machine spindle. Note that the opposed taper roller bearings are assembled in close proximity at one end only of the shaft. Therefore, if the shaft heats up and expands when running, it is free to slide axially in the parallel bearings supporting the opposite end.

Material requirements

(a) *Wear prevention.* For wear prevention in rolling bearing materials it is important that they have a high degree of hardness and are finished to a high degree of smoothness.

(b) *Fatigue resistance.* The elements of rolling bearings are constantly flexing at the point of contact (rather like a tyre on the road) and this leads to high levels of fatigue stress. Therefore a material with high fatigue resistance is essential and it must be finished without surface imperfections which might accelerate fatigue failure (see Sections 9.2 (surface fatigue) and 13.3).

(c) *Strength.* At one time case hardened alloys were used exclusively for rolling bearings as it was thought that this would provide the necessary hardness to prevent wear and, at the same time, would provide the toughness and strength necessary to support the load and provide a degree of deformation to assist alignment. In practice the constant flexing (elastic deformation) caused the hard case to flake away from the more flexible core. Modern practice for high quality bearings favours the use of high carbon-chrome steel alloys, hardened

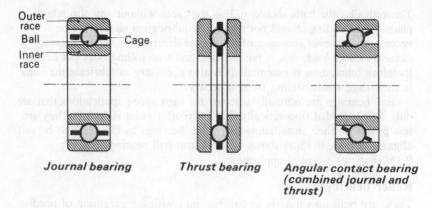

Journal bearing **Thrust bearing** **Angular contact bearing (combined journal and thrust)**

(a) Typical ball bearings

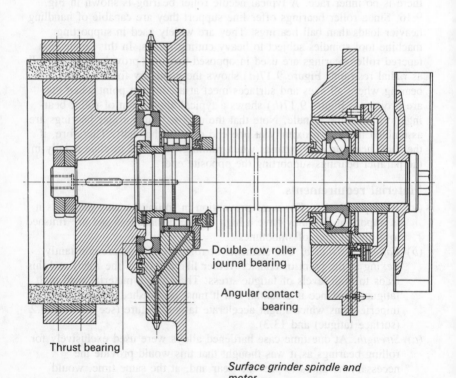

Double row roller journal bearing

Angular contact bearing

Thrust bearing

Surface grinder spindle and motor

(b) Applications of ball bearings

Fig. 9.15 Ball bearings

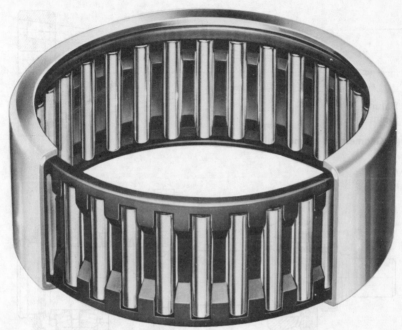

Fig. 9.16 Needle roller bearing

to give a martensitic structure throughout, except where heavy shock loads are encountered for which applications case-hardening alloys are still preferred.

9.13   Bearing materials (metallic)

Traditionally, sliding bearing materials were alloys of tin, lead and antimony referred to as 'white-metal' alloys or 'Babbitt-metal'. Copper base alloys, aluminium base alloys and polymeric materials are also used nowadays for sliding bearings.

White-metal alloys

These alloys represent the majority of bearing metals in which small particles of a hard phase are embedded in a ductile solid solution matrix. The hard particles are usually intermetallic compounds which have good wear resistance and anti-friction properties. The ductile matrix provides toughness (shock resistance) and sufficient ductility to allow the bearing to 'bed in' during the running-in period.

White-metal alloys may be either tin based or lead based. High quality Babbitt metals may contain up to 90 per cent tin and up to 10 per cent antimony. Typical alloys are listed in Table 9.1. Tin and antimony form

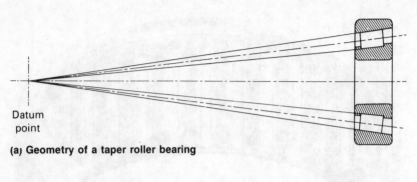

(a) **Geometry of a taper roller bearing**

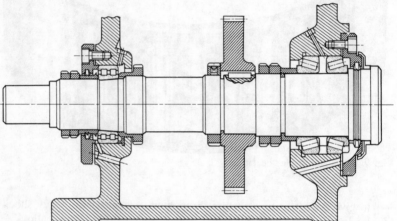

(b) A machine-tool spindle

Fig. 9.17 Roller bearings

cuboid crystals of the tin-antimony intermetallic compound SbSn. These cuboids of the intermetallic compound constitute the hard, anti-friction phase. In a lead-free Babbitt metal these hard cuboids are dispersed through a soft, ductile matrix of a solid solution of antimony in tin.

Lead is added to white metals to reduce the cost and to increase the plasticity of the material to ease alignment. In such alloys there is a tendency for the intermetallic tin-antimony cuboids to float to the surface of the molten metal. The molten metal eventually solidifies to form a eutectic of two solid solutions. One of these will be lead rich and one will be tin rich. To prevent the segregation of the tin-antimony cuboids, copper is added. The copper forms an intermetallic compound with the tin which separates out before the tin-antimony cuboids which are trapped in the mesh of the copper-tin matrix thus preventing segregation.

Table 9.1 Some typical bearing materials

Category	Composition (%)					Properties and applications
	Sn	Sb	Cu	Pb	P	
White metal	93	3·5	3·5	—	—	Big-end bearings for light and medium duty, high-speed internal combustion engines
	86	10·5	3·5	—	—	Main bearings for light and medium duty, high-speed internal combustion engines
	80	11·0	3·0	6·0	—	General purpose, heavy duty bearings. Lead improves plasticity where alignment is a problem
	60	10·0	28·5	1·5	—	Heavy duty marine reciprocating engines, electrical machines
	40	10·0	1·5	48·5	—	Low cost, general purpose, medium duty, bearing alloy
Bronze	10·5	—	89	—	0·5	Good anti-friction properties, suitable for heavy loads, rigid
	10·0	—	79·9	10	0·1	Good anti-friction properties, lubrication not critical, lead content reduces rigidity and helps alignment
	3	—	74	23	—	Leaded (plastic) bronze, excellent self-alignment properties due to high lead content. For duty intermediate between white metal and phosphor bronze
	Fe	C	Si	Mn	S/P	
Cast Iron	94	3·3	1·3	1·0	0·1/0.3	The flakes of graphite (carbon) in grey cast iron gives it self lubricating properties. Suitable for heavy duty, low-speed applications where lubrication is difficult, e.g. machine tool slideways
Plastic	Polytetrafluoroethylene					**Teflon :** Can withstand much higher temperatures than most plastics. Very expensive anti-friction coating — very low coefficient of friction. Does not require lubrication
	Polyamide					**Nylon :** Can be moulded into bushes and gears. Does not require lubrication. Use for office and food processing machinery
	High density polyethylene					Low cost bearings. Does not require lubrication. Cannot support such high loads as Nylon or Teflon

Sn = Tin, Sb = Antimony, Cu = Copper, Pb = Lead, P = Phosphorus, Fe = Iron, C = Carbon, Si = Silicon, Mn = Manganese, S = Sulphur

Copper-lead alloys

These alloys produce a bearing material with a hard phase matrix within which there is a softer phase network. They owe their anti-friction properties to the smearing of a thin film of lead over the surface of the harder copper during the 'running-in' period. Shearing of the asperities occurs in this lead film, whereas in Babbitt metals it occurs in the soft matrix. An example of a copper lead alloy is given in Table 9.1.

Both Babbitt and copper-lead alloys are weak and have to be coated onto a stronger bearing shell. This is often steel and, to prevent damage if the white-metal layer is worn through, an intermediate layer of a strong bearing metal such as a tin-bronze is provided.

Aluminium base bearing materials

Bearing materials based upon aluminium-silicon alloys have been known for some time and were developed originally for heavy duty applications in aircraft engines (piston type). They did not become a commercial proposition until the introduction of roll bonding techniques for making bi-metal strips. The aluminium-silicon-tin alloy is roll bonded onto a steel shell. The aluminium-silicon alloy usually contains 7 per cent tin to prevent galling and to improve the anti-friction properties. Small amounts of nickel and copper may also be present. Filtration of the lubricating system is very important with such bearing materials as they are very hard and have poor *embeddability* (see Section 9.15), and any foreign matter in the bearing will score the mating surfaces instead of becoming harmlessly embedded in the soft matrix below the surface.

Copper base alloys

Tin-bronzes, such as gunmetal and phosphor bronze, and aluminium bronzes are widely used for heavy duty bearing materials. They are stronger than the materials so far described and are self-supporting. These bearing materials are very rigid and careful machining and alignment is essential since very little bedding-in can take place. Lead is added (leaded-bronze) to improve the plasticity and anti-friction properties of the alloy, but with some loss of strength. Typical alloys, their composition and some typical applications are listed in Table 9.1.

Porous bearings

These are produced by sintered powder metallurgy processes. The bronze bearing metal alloy in powder form is compressed into the rquired shape and this compact is then sintered at a high temperature in a reducing atmosphere to produce a micro-porous structure (a metal 'sponge'). Typical structures are shown in Fig. 9.18. Bearings produced in this manner are impregnated with a suitable lubricant which normally lasts the life of the bearing. They are used for applications, such as domestic appliances where regular maintenance and lubrication cannot be relied upon, or for applications where an automatic lubrication system would be inconvenient or excessively expensive. Alternatively, provision may be

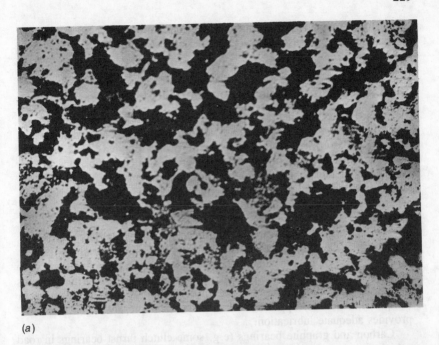

(a)

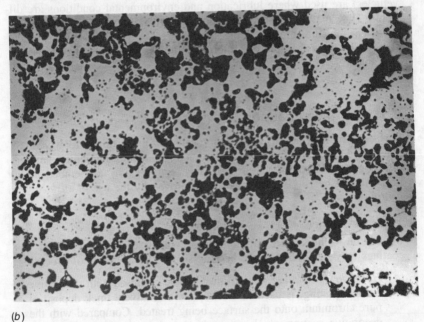

(b)

Fig. 9.18 (a) Microstructure of low density high porosity metal (b) Microstructure of high density low porosity metal

made for periodical replenishment of the lubricant when the use of a porous bearing increases the working period between servicing.

9.14 Bearing materials (non-metallic)

Polymeric materials such as teflon, nylon and high-density polystyrene can all be used for sliding bearings when lightly loaded and when the speed is not excessive so that temperature rise is minimal. They have the advantage that normal atmospheric moisture is sufficient lubricant and no oil or similar lubricant is required. This makes them ideal for use in office machinery and food-processing equipment. They are also used where corrosive environments preclude the use of metallic bearing materials.

For heavier duty applications reinforced polymer materials such as 'Tufnol' may be used successfully. For example, this material has largely ousted the use of lignum vitae wooden inserts for the propeller shaft stern bearings of large ships. The presence of sea-water in the bearing prevents the use of metallic materials because of corrosion. A further advantage in using Tufnol for this application is that the water present in the bearing provides adequate lubrication.

Carbon and graphite bearings (e.g. some clutch thrust bearings in road vehicles) are used where lubrication and environmental conditions are difficult. They may or may not be lubricated with water. Usually there is sufficient moisture present in the atmosphere to provide adequate lubrication. Some typical examples of non-metallic bearing materials are listed in Table 9.1.

9.15 Surface coatings

Surface hardening has already been considered in Section 5.13 which considered case hardening, flame-hardening, induction hardening, and nitriding. In addition, the wear resistance, anti-friction, and embeddability properties of bearing surfaces can be improved by the use of *surface coatings*.

Wear resistance

In addition to the surface hardening processes described above, hard coatings may be applied to the bearing surface by a variety of processes.

(a) *Hard-chrome plating*, which must not be confused with decorative chrome plating, is used to electroplate a relatively thick deposit of pure chromium onto the surface being treated. Compared with the decorative coating which may only be a few microns thick (1 micron = 1 μm), a hard chrome deposit may be up to 0.4 mm thick. After finishing by grinding and superfinishing (e.g. honing) such a surface has high hardness, very low friction and high corrosion resistance. The slight porosity of the surface can be carefully controlled to improve its lubrication properties.

(b) *Hard metals*, such as stellite — a cobalt based alloy which is so hard that it can only be machined by grinding — can be deposited on the bearing surface by welding using an oxy-acetylene torch with a 'reducing' flame setting. A considerable thickness of metal can be deposited and such a surface not only has considerable wear resistance, it can operate continuously at high temperatures.

(c) *Ceramics and carbides*, can be deposited by various processes to provide wear resistant bearing surfaces. Ceramics such as alumina or zirconia may be sprayed onto the surface using a 'gun' in which the powder is melted by an oxy-acetylene flame and the molten material is sprayed onto the surface by compressed air. Hard carbides, such as tungsten carbide, may be applied in a similar manner or they may be preformed and brazed onto the bearing surface. For sliding bearings a material with hard carbides dispersed through a softer and tougher matrix is to be preferred, for example tungsten carbide particles in a titanium carbide matrix.

Embeddability

The development of harder, heavy-duty bearing materials such as aluminium-silicon alloys and the lead bronzes presents problems resulting from their low embeddability properties, that is, they lack the ability to absorb small, abrasive particles of foreign matter and this can result in damage to the shaft journals. The problem is overcome by *surface coating* these bearing materials with an *overlay* of lead-indium or lead-tin alloy. This overlay is plated onto the bearing metal and is just thick enough to absorb the dirt particles, but not so thick as to reduce the strength of the bearing. Such an overlay is shown in Fig. 9.19.

Friction

The use of teflon for solid plain bearings has already been considered. It is also used as a surface coating on machine tool slideways where friction and the problems associated with 'stick-slip' has to be reduced to a minimum, for example in computer controlled machine tools where excessive 'stick-slip' could adversely affect accurate positioning of the work table.

Phosphite coatings also have a low coefficient of friction and they are frequently used to coat steel sheet and steel wire before extreme cold drawing processes.

9.16 Strength of sliding bearing materials

An important factor which must be taken into account when assessing the suitability of a sliding bearing material is that the effective strength of a

Steel particles embedded in a lead–indium overlay. Fig. 9.19

Fig. 9.19 Embeddability. The bearing material is cast leaded bronze (designated VP2) and is electroplated with a lead-indium overlay. The dark, central debris particle is composed of silica, while the two light-coloured particles are ferrous (i.e. either steel of cast iron).

233

thin layer of material increases rapidly as its thickness decreases. This is shown for a typical Babbitt metal in Fig. 9.20. For example, *micro-Babbitt* (Babbitt metal bearings less than 0.01 mm thick) layers are substantially stronger than the more normal thickness of 0.4 mm. Again, plated overlays of very soft alloys such as lead-indium and lead-tin can be used over a copper-lead intermediate strata without reducing the strength of the bearing. The plating thickness, in this instance, has to be

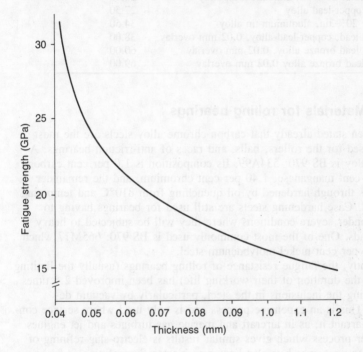

Fig. 9.20 Relationship between effective strength and thickness for a typical thin-wall bearing material

a composite between strength (which requires a thin film) and embeddability (which requires a rather thicker film). Table 9.2 lists some of the thin-wall materials in current use and the minimum load they can be expected to withstand for, say, a medium duty internal combustion engine.

Finally, the bearing material must also be compatible with the shaft material with which it is used. This is a complex subject beyond the scope of this book. However, data is available from bearing manufacturers for combinations of materials which have been proved to be satisfactory by prolonged experimental research and practical experience.

Table 9.2 Safe loads for thin-wall bearings

Bearing material (thin-wall)	Safe load (GPa)
Lead-tin based Babbitt metal	12.50
Micro-Babbitt metal	17.50
Non-recticular 20% tin, aluminium-tin alloy	24.15
Sintered copper-lead alloy	27.50
Recticular 20% tin, aluminium-tin alloy	34.50
Cast 27% lead, copper-lead alloy, 0.02 mm overlay	38.00
Cast 23% lead bronze alloy, 0.02 mm overlay	60.00
Cast 5% lead bronze alloy 0.02 mm overlay	69.00

9.17 Materials for rolling bearings

It has been stated already that carbon-chrome alloy steels are the most widely used for the rollers, balls, and races of anti-friction bearings. A typical alloy is BS 970: 534A99. Its composition is 1.0 per cent carbon, 0.45 per cent manganese, 1.40 per cent chromium, and the remainder iron. It is through-hardened by oil quenching from 810°C and tempering at 150°C. Case hardening steels are still used for bearings having to operate under severe conditions where they will be subjected to heavy shock loads. One of the most commonly used is BS 970: 665M17 which is a 1.75 per cent nickel-molybdenum steel.

Recently, the fatigue resistance of rolling bearings (usually the limiting factor in the duration of their working life) has been improved 2.5 times by lowering the inclusions in the steel, particularly by vacuum de-gassing. This is an expensive process and is only used where service conditions warrant it, as in aircraft and marine gas turbines and jet engines. A cheaper process which gives similar results is electro-slag-refining of the steel (e.s.r.), which is also known as electro-flux-refining (e.f.r.). Special steels have been used for corrosive environments and high-temperature environments. These tend to be less hard and the life of the bearing suffers accordingly. The normal maximum operating temperature for a rolling bearing is 150°C. In any case, there are lubrication problems above this temperature.

10 Shaping materials

10.1 Casting

Metals can be formed directly from the molten state by pouring them into moulds and allowing them to cool and solidify. Obviously, the mould must be made from a material with a higher melting point than that of the molten metal from which the casting is to be made. The mould also contains a cavity in the form of the finished product into which the molten metal is poured. The form of this cavity is determined by ramming sand round a wooden or GRP pattern. The pattern is the same shape as the finished casting but slightly larger to allow for shrinkage. After ramming, the mould is opened so that the pattern can be removed from the cavity. The mould is then re-assembled ready for pouring.

One of the most widely used materials for making moulds is moulding sand, and sand moulds are used for most high melting point materials such as cast iron. The moulds can be made individually by hand where small quantities are required or by semi- and fully-automated processes where they are required for quantity production. Figure 10.1 shows a section through a typical two part sand mould. The molten metal is poured from a ladle into the *runner* and the air displaced from the cavity by the molten metal escapes through the *risers*. There must be a riser above each high point of the cavity to prevent air locks. Pouring continues until the molten metal is seen at the top of each riser. This ensures that the mould cavity is full. It also provides surplus metal which can be drawn back into the mould as shrinkage takes place during cooling. This avoids shrinkage cavities occurring in the casting. The mould must also contain *vents*. These are fine holes made with a wire after the mould is

236

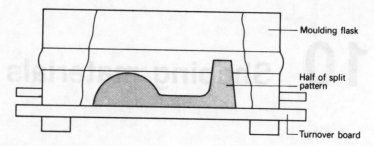

(a) **Half pattern in position**

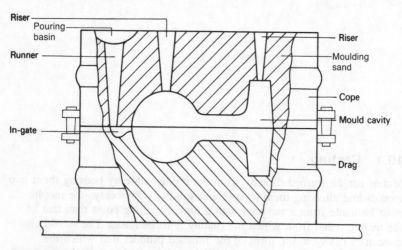

(b) **Complete mould — pattern removed**

Fig. 10.1 Two-part sand mould

complete. The holes stop just short of the mould cavity. The purpose of the vents is to release steam and other gases which are generated when the hot metal comes into contact with the moist moulding sand. (Moisture is required in moulding sand so that it will bind together and keep its shape.) If the mould was not vented, then the release of steam and gases would cause bubbles to collect in the casting causing 'blow holes' and 'porosity'. When the metal has solidified to form the casting the mould is broken open and the casting is removed. The runners and risers are cut off and melted down again and the casting is ready for machining.

10.2 Shrinkage and machining allowance

It has already been stated that the pattern has to be made oversize to allow for shrinkage of the metal as it cools. This is called the *shrinkage*

Table 10.1 Shrinkage allowance

Material	Shrinkage allowance
Aluminium	21·3 mm/m
Brass	16·0 mm/m
Cast iron	10·5 mm/m
Steel	16·0 mm/m

allowance. Table 10.1 lists some common metals and the magnitude of the shrinkage allowance. The pattern maker does not have to calculate this allowance but uses a special rule which is already engraved oversize. Obviously a different rule or scale has to be used for each type of metal being cast. In addition to shrinkage allowance, the pattern must also be made oversize wherever the casting is to be machined, that is, a *machining allowance* has to be superimposed on top of the shrinkage allowance. Not only must sufficient additional metal be provided to ensure that the casting 'cleans up', but sufficient metal must be provided to ensure that the tip of the cutting tool operates well below the hard and abrasive skin at the surface of the casting. This is shown in Fig. 10.2.

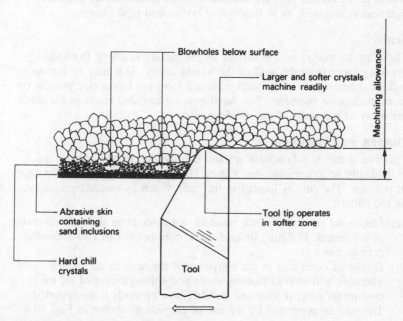

Fig. 10.2 Machining allowance

10.3 Casting defects

The mechanism by which metals solidify and the effect this has upon their grain structures and properties has already been discussed in Sections 2.11 and 2.12. The practical defects which occur in the foundry can be listed as follows.

Blowholes

These are smooth round holes with a shiny surface usually occurring just below the surface of the casting. Because they are not normally visible until the casting is machined, their presence can mean scrapping a casting on which costly machining has already been carried out. Blowholes are caused by steam and gases being trapped in the mould (Section 10.1). This may result from inadequate venting, incorrectly placed risers, excessive moisture in the sand or excessive ramming reducing the permeability of the sand. Permeability is the ability of the sand to allow entrapped gases to escape between the individual sand particles.

Porosity

This is also caused by inadequate venting, but in this instance the trapped gases do not form large bubbles near the surface of the casting. Instead, a mass of pinpoint bubbles are spread throughout the casting rendering it porous and 'spongy'. Inadequate 'de-gassing' of the metal immediately prior to pouring is a frequent cause of porosity. Apart from being a source of weakness, porosity renders castings useless where pressure tightness is required, as in fluid valve bodies and pipe fittings.

Scabs

These are blemishes on the surface of the casting resulting from sand breaking away from the wall of the mould cavity. This may be due to lack of cohesiveness in the sand resulting from too low a clay content or from inadequate ramming. Too rapid pouring can also result in the scouring away of the walls of the mould cavity.

Uneven wall thickness

This can occur in hollow castings and may be due to two causes, either individually or in combination. Figure 10.3 shows a mould with the core in position. The core is located in the 'prints' left by suitable projections on the pattern.

(a) *Misplaced cores* due to the moulder not assembling the core correctly in the mould. Ill-fitting or inadequate core-prints can also allow the core to move.

(b) *Displaced cores* due to the buoyancy of the sand in the molten metal. The core will tend to float upwards and although trapped in, the core-prints may, if long and slender, bow upwards if unsupported. This can be prevented by the use of chaplets as shown in Fig. 10.4

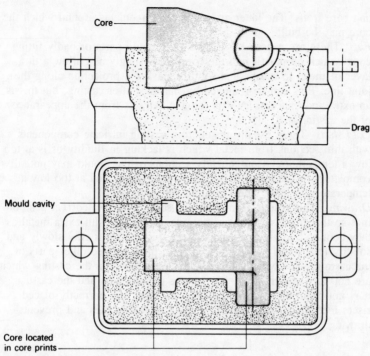

Fig. 10.3 Use of cores

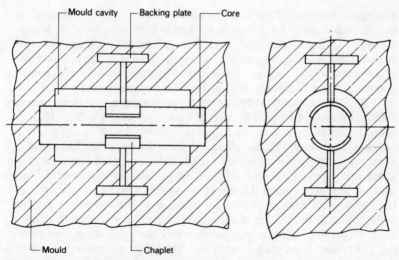

Use of chaplets to support a core

Fig. 10.4 Use of chaplets to support a core

and core irons. The latter are metal reinforcements around which the core may be built.

(c) *Fins*. These are caused by badly fitting mould parts or badly fitting cores which do not fit into the core-prints snugly and allow a thin layer of metal to escape and solidify leaving a projection along the joint line. Fins can be removed by 'fettling' after casting, but this is an extra and unnecessary expense and detracts from the appearance of the casting.

(d) *Cold shuts*. These usually result from casting intricate components with thin sections from metal which is lacking in fluidity or is at too low a temperature. Consequently sections of the mould may not fill completely or the metal may flow too sluggishly and at too low a temperature to unite when separate streams meet.

(e) *Drawing*. This results from lack of risers, or their incorrect positioning, so that thick sections are not adequately fed with extra metal as they cool and contract. Further, since thick sections cool slowly and solidify last, metal may be drawn from them to feed other parts of the casting. This will result in holes and hollows in the casting which are not only unsightly but are sources of weakness and the casting may not clean up when machined. Sufficient and correctly placed risers feed metal back into the casting as it solidifies and prevents drawing.

10.4 Plastic deformation

The basic principles of the hot-working, cold-working and recrystallisation of metals in the solid state have already been introduced in Section 5.1. Metals, in the solid state, have a crystalline structure and the crystals or grains are made up of particles arranged in strict geometric patterns called space lattices (Section 2.8 *et seq.*). It is due to the strict geometric symmetry of these space lattices that plastic deformation can occur in the solid state. When plastic deformation occurs during hot-working or cold-working, planes of atoms slip past each other as shown in Fig. 10.5(*a*). These planes of movement are called *slip planes*. Usually, slip planes lie between, and parallel to, the planes of greatest atomic density as shown in Fig. 10.5(*b*). When a ductile or a malleable crystalline material is subjected to an applied force of sufficient magnitude, movement of the lattice structure can occur along the slip planes within a crystal as shown in Fig. 10.6. Obviously movement does not occur in all the slip planes available in the material, but only in those planes which are at a suitable angle to the applied force. Slip can only occur where the grain is not constrained, and slip cannot cross grain boundaries. Thus slip can only occur within a grain; the bigger the grain, the greater the number of slip planes available, and the greater the amount of slip which can take place. This is borne out in practice, since fine grain materials are generally less

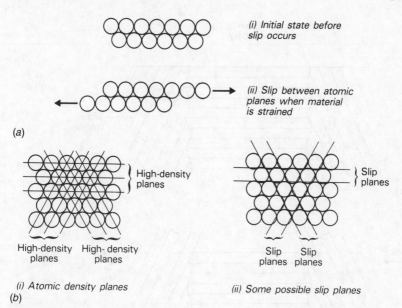

(a)

(i) Initial state before slip occurs

(ii) Slip between atomic planes when material is strained

(b)

High-density planes

High-density planes   High-density planes

(i) Atomic density planes

Slip planes

Slip planes   Slip planes

(ii) Some possible slip planes

Fig. 10.5 Slip planes (a) Slip between planes of high atomic density (b) The orientation of slip planes

ductile and malleable than the same materials after processing by heat treatment to increase the grain size.

Initally, it was thought that slip occurred as the result of one block of atoms moving simultaneously over another, rather like sliding a carpet bodily along a floor as shown in Fig. 10.7(a). However, calculations on the strength of metals proved that this was not true, and that slip occurs through a system of *dislocations* rather like moving a carpet a little at a time by bunching it as shown in Fig. 10.7(b) and moving the 'bunch' along progressively. Similarly, slip can be assumed to take place progressively by the movement of a dislocation along the slip planes. This dislocation is a misfit in the geometry of the crystal lattice and it separates the region that has been subjected to slip from regions where slip has not occurred. In this manner, the dislocation can travel across the crystal with little applied force (far less than would be required to cause the slip of a block of atoms simultaneously). When the dislocation has travelled across the crystal the whole plane has slipped one atomic distance (approximately 4×10^{-10} mm). This is obviously a very small distance, and a measurable plastic strain involves the movement of a large number of dislocations. The mechanism of a dislocation is shown in Fig. 10.8.

Another mechanism by which deformation can take place is *twinning*. The principle of deformation by twinning is shown in Fig. 10.9. Unlike

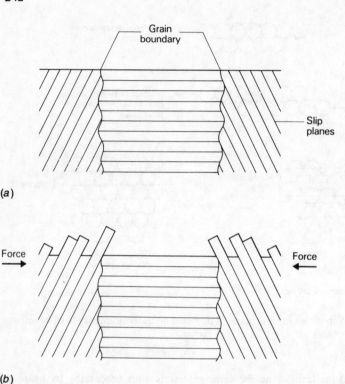

Fig. 10.6 Formation of slip bands during the plastic deformation of a metal (*a*) Slip planes before the application of a force (*b*) Movement of slip planes after the application of a force

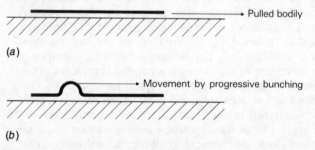

Fig. 10.7 'Carpet analogy' of dislocation

slip, where all the atoms in a block move the same distance, in twinning each successive plane of atoms in a block moves a different distance. When twinning is complete, the deformation of the crystal lattice will result in one half of the twin becoming the mirror image of the other half as shown in Fig. 10.9(*b*). Like slip, twinning proceeds by a series of dislocations. The force required to cause twinning is generally greater

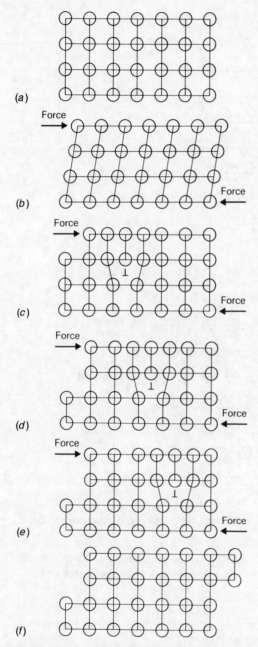

Fig. 10.8 Principle of dislocation (*a*) Unstressed crystal lattice (*b*) Elastic deformation. Lattice is distorted but slip has not yet occurred (*c*) Plastic deformation. Dislocation commences at '⊥' (*d*) Dislocation moves across lattice (*e*) Dislocation moves across lattice (*f*) Deformation complete. Dislocation moves out of the crystal to form a slip band

244

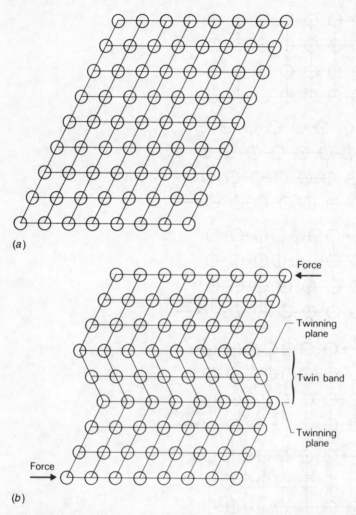

(a)

Force

Twinning plane

Twin band

Twinning plane

Force

(b)

Fig. 10.9 Principle of twinning (a) Crystal lattice prior to deformation (b) Crystal lattice after deformation

than the force to cause slip. Twinning is not as common as block slip and occurs mainly when metals are shock loaded at low temperatures (e.g. cold-heading rivets).

10.5 Cold-working

Figure 10.10(a) shows the effect of drawing a metal rod through a die so as to reduce its diameter and increase its length. As it passes through the die, the metal undergoes severe plastic deformation and the equi-axed

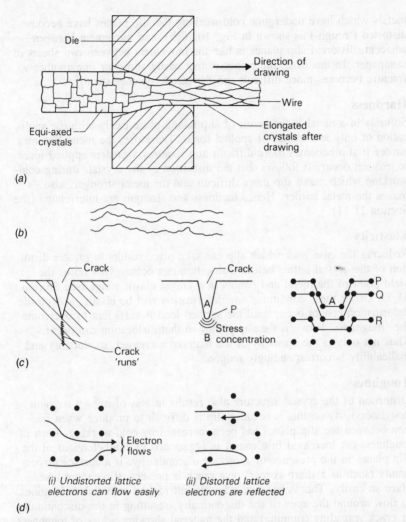

Fig. 10.10 Some effects of cold-working metals (*a*) Effect of cold-working on crystal structure (*b*) Distorted slip planes in a cold-worked metal (*c*) Mechanism of crack propagation (*d*) Effect of cold-working on electrical conductivity

crystals typical of a metal in the annealed condition become elongated and distorted. In cold-working this distortion occurs below the temperature of recrystallisation (see Section 5.1), and the crystals remain in the distorted condition. This affects the properties of the metal as follows.

Strength

The tensile strength of the metal is increased. This is because metals only fracture by movement of the lattice along the slip planes. In the case of metals which have undergone cold-working the slip planes have become distorted ('rough') as shown in Fig. 10.10(b), and movement between adjacent, distorted slip planes is like the movement between two sheets of sandpaper. In this case, the planes cannot easily slip over one another so fracture becomes more difficult and the metal becomes stronger.

Hardness

Softness in a metal is the result of slip taking place easily with the application of only low values of applied force. Therefore, the metal becomes harder if slip becomes more difficult and requires a greater applied force to make it occur. It follows that the distortion of the crystals during cold-working which make slip more difficult and the metal stronger, also makes the metal harder. Hence hardness and strength are interrelated (see Section 11.11).

Elasticity

Reducing the ease with which slip can take place results in greater distortion of the crystal lattice before dislocation can occur. This raises the yield point of the metal and results in a longer elastic range (see Section 11.3). Under these conditions any deformation will be elastic, and plastic deformation cannot occur until the applied load is sufficient to overcome the 'roughness' between the slip planes so that dislocation can occur. Thus not only is the elasticity of the material increased, its ductility and malleability is correspondingly reduced.

Toughness

Distortion of the crystal structure also results in loss of impact strength (toughness). Again this is because slip is difficult to produce when friction between the slip planes has been increased by cold-working. Lack of toughness (or increased brittleness) is aggravated by the behaviour of the slip planes in the presence of a surface discontinuity. If a surface discontinuity (such as a sharp corner) or a crack is present, slip cannot take place so easily. This is because it is difficult for the essential dislocations to flow around the apex of the discontinuity, resulting in the discontinuity or crack spreading (running) and the material showing a loss of toughness and a corresponding increase in brittleness. This is why impact (toughness) test specimens are usually notched. Figure 10.10(c) shows the mechanism of crack propagation in a brittle metal. It makes no difference whether the brittleness has been imparted by crystal distortion during cold-working or whether the brittleness has been imparted by heat treatment (quench-hardening).

At the base of the discontinuity the stress will be large when the metal is in tension, that is, there will be a stress concentration at the point P. At the base of the discontinuity the plane P tries to slip over the plane Q and prevent the discontinuity from running and spreading. If slip cannot easily take place, then the stress at P will cause the intermolecular bonds at Q to break and transfer the stress to the plane R, and so on. Thus the crack spreads and the metal is said to be brittle.

Machinability

This is improved by cold-working since the metal becomes stiffer, resulting in a cleaner shear at the point of cutting and a correspondingly improved surface finish.

Electrical conductivity

Cold-worked metals have a lower electrical conductivity than annealed metals. This is because cold-working distorts the whole crystal lattice and makes it more difficult for electron flow to occur as shown in Fig. 10.10(d).

10.6   The heat treatment of cold-worked metals

If the working of the metal is excessive it will work-harden to such an extent that its increased brittleness will result in the metal breaking under reduced load conditions. This is precisely what happens during a tensile test to destruction (see Section 11.2). The properties imparted to the metal by cold-working will remain indefinitely until the structure of the metal is changed by heat treatment.

The heat treatment process will depend upon the composition of the material and the severity of the cold-working it has received. The process annealing of plain-carbon steels has been described in Section 5.3, and the solution treatment of the heat-treatable non-ferrous alloys was described in Section 7.5. These processes are not appropriate to pure metals such as aluminium and copper and non-ferrous alloys such as brass which do not respond to solution and precipitation treatments.

Recovery

The minimum heat treatment for a cold-worked metal is simple stress relief. This occurs at quite low temperatures, particularly for non-ferrous metals and alloys. There is no change in grain structure, but individual atoms move within the crystal lattice. This effect is called *recovery* and occurs at a temperature of about one third the melting point of the metal when calculated on the Kelvin temperature scale.

Example 10.1 Estimate the recovery temperature for a mild steel pressing if its melting temperature is taken as 1520°C.

The melting point for the steel	= 1520°C
	or 1520 + 273
	= 1793 K
Thus the recovery temperature	= 1793 × 1/3
	= 598 K (approx)
	or 598 − 273
	= 325°C

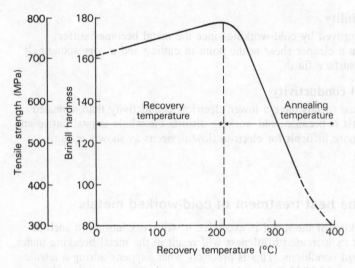

Fig. 10.11 The effect of heat treatment on cold-worked 70/30 brass

Treatment at the recovery temperature does not adversely affect the hardness and strength resulting from cold-working and may even enhance these properties in some instances as shown in Fig. 10.11. Stress relief annealing of 70/30 brass after it has been severely cold-worked prevents 'season cracking' occurring later in its service life.

Recrystallisation

The recrystallisation of metals has been described previously in Section 5.1. If cold-worked metal is to undergo further plastic deformation, then it must be heat treated to restore the grain structure to the annealed condition by *recrystallisation*. The recrystallisation temperature depends largely upon the degree of cold-working which the material has undergone prior to heat treatment. The more severe the cold-working, the lower will be the recrystallisation temperature. The recrystallisation temperature is generally lower for pure metals than it is for alloys. For example, as shown in Tables 7.8 and 7.9, the addition of 0.5 per cent of the metal arsenic to copper produced an alloy which would sustain its strength and hardness at elevated temperatures. This is due to the presence of arsenic in the alloy raising the recrystallisation temperature from about 200°C for pure copper to 500°C for the arsenical copper alloy.

Just as the recovery temperature can be approximated as one third the melting temperature as calculated on the Kelvin scale, so the recrystallisation temperature can be approximated as one half the melting temperature of the metal or alloy as calculated on the Kelvin scale. Some metals, such as copper, recrystallise at a slightly lower temperature since they can sus-

Example 10.2 Estimate the recrystallisation temperature for a mild steel pressing if its melting temperature is taken as 1520°C.

$$
\begin{aligned}
\text{The melting point for mild steel} \quad &= 1520°C \\
&\text{or } 1520 + 273 \\
&= 1793 \text{ K}
\end{aligned}
$$

$$
\begin{aligned}
\text{Thus the recrystallisation temperature} &= 1793 \times \tfrac{1}{2} \\
&= 896.5 \text{ K} \\
&\text{or } 896.5 - 273 \\
&= \underline{\underline{623.5°C}}
\end{aligned}
$$

tain very severe cold-working and crystal deformation prior to heat treatment.

Reference back to Section 5.3 for sub-critically annealing mild steel shows that this approximation gives an acceptable recrystallisation temperature. Some metals such as lead and tin recrystallise at temperatures below room temperature and it is virtually impossible to cold-work such metals under normal working conditions.

Grain growth

Finally, the effect of crystal structure on grain growth must be considered. Figure 10.12 shows the relationship between grain size after recrystallisation annealing and the amount of deformation received prior to annealing. If the metal is only lightly worked (5 to 10 per cent reduction) the crystals will be, correspondingly, only slightly distorted with

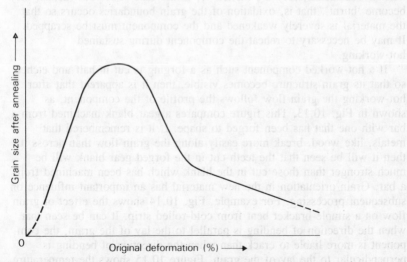

Fig. 10.12 Relationship between grain size and deformation

few stress points. Therefore upon annealing, few nuclei will be created and these will have room to grow into large, coarse grains. This coarse grain structure leads to loss of strength and the formation of an 'orange peel' surface. Therefore, if the metal is only lightly worked, heat treatment should be restricted to 'recovery' treatment. Recrystallisation annealing should be reserved for severely cold-worked metals if a fine grain is to be achieved. It must be remembered that for most metals and alloys other than the ferrous metals, the heat-treatable aluminium alloys and some aluminium bronzes, cold-working is the only way in which the metals may be hardened, and that the subsequent recrystallisation is the only way in which grain refinement can be achieved.

10.7   Hot-working

This has already been described as the forming of metals by plastic flow above the temperature of recrystallisation. If the temperature of the metal being worked is sufficiently high, recrystallisation takes place as quickly as the crystals become deformed and the metals can be heavily worked with ease and without risk of cracking. As the temperature falls during processing, recrystallisation occurs more slowly. Not only is more force required to achieve plastic deformation, but there is an increased risk of surface cracks appearing. A hot-working process should be matched to the heating and cooling cycle of the component so that the process is completed at a temperature sufficiently above its recrystallisation temperature that grain growth occurs. Care must be taken not to raise the temperature of the component too high initially so that, in the case of non-ferrous metals, they are melted or, in the case of ferrous metals, they become 'burnt', that is, oxidation of the grain boundaries occurs so that the material is severely weakened and the component must be scrapped. It may be necessary to reheat the component during sustained hot-working.

If a hot-worked component such as a forging is cut in half and etched so that its grain structure becomes visible, then it is apparent that after hot-working the grain flow follows the profile of the component, as shown in Fig. 10.13. This figure compares a gear blank machined from a bar with one that has been forged to shape. If it is remembered that metals, like wood, break more easily along the grain flow than across it, then it will be seen that the teeth cut in the forged gear blank will be much stronger than those cut in the blank which has been machined from a bar. Grain orientation in the new material has an important influence on subsequent processing. For example, Fig. 10.14 shows the effect of grain flow on a simple bracket bent from cold-rolled strip. It can be seen that when the direction of bending is parallel to the lay of the grain, the component is more liable to crack than when the direction of bending is perpendicular to the lay of the grain. Figure 10.15 shows the temperature ranges for the hot-working of some typical metals.

251

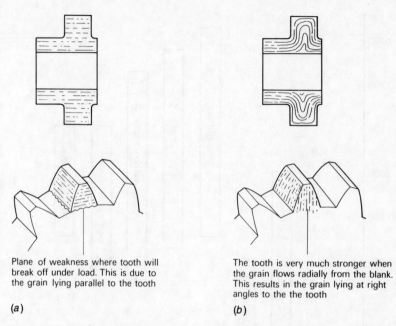

Plane of weakness where tooth will break off under load. This is due to the grain lying parallel to the tooth

(a)

The tooth is very much stronger when the grain flows radially from the blank. This results in the grain lying at right angles to the the tooth

(b)

Fig. 10.13 Grain orientation (a) Machined from bar (b) Machined from forging

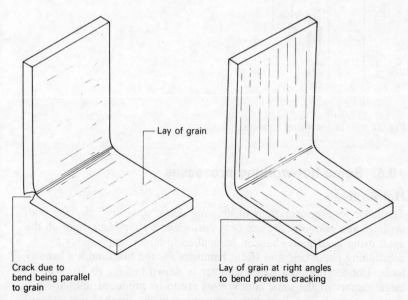

Lay of grain

Crack due to bend being parallel to grain

Lay of grain at right angles to bend prevents cracking

Fig. 10.14 Effect of grain flow on subsequent processing

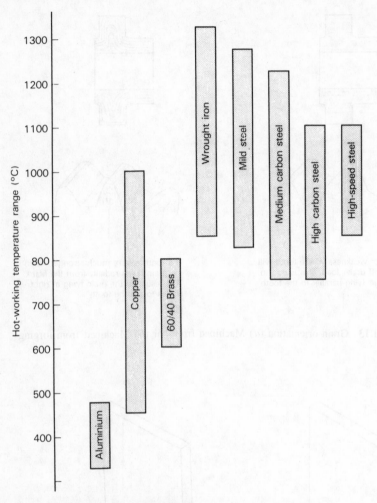

Fig. 10.15 Hot-working temperatures

10.8 Some hot-working processes

Hot-forging

The basic forging operations of *drawing down, upsetting, piercing and drifting,* and *swaging,* are not only performed by the blacksmith on the anvil using hand tools, but can be applied to larger components by substituting pneumatic and steam hammers for the blacksmith's hand tools. Hot-forging on a steam hammer is shown in Fig. 10.16. Where a large number of the same components are to be produced, closed-die forging is used. The die cavity is the shape of the finished component and a section through such a die is shown in Fig. 10.17. A set of forging dies for use with a drophammer or 'stamp' is shown in Fig. 10.18. For very large forgings hydraulic presses are used, as their slow but steady squeeze ensures that grain flow occurs right to the centre of the work, whereas the sharp blow of the hammer only produces surface flow in any but small components.

Fig. 10.16 Forging — steam hammer

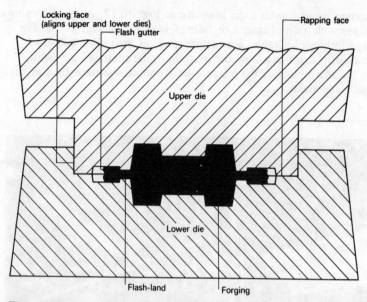

Fig. 10.17 Closed-forging die

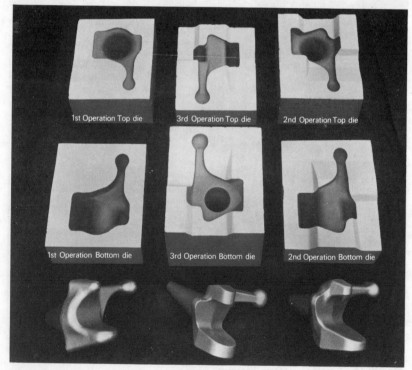

Fig. 10.18 Drop-forging

Hot-rolling

The hot rolling of an ingot into a slab is shown in Fig. 10.19. The white-hot slab is just leaving the rolls and is supported on a motorised roller bed. The slab passes backwards and forwards between the work rolls of the mill and is gradually reduced in thickness. Reference back to Fig. 2.15 shows the structure of a typical cast ingot prior to hot-rolling. The structure changes from fine chill crystals at the surface to coarse equi-axed crystals at the centre, resulting from the different rates of cooling. In addition there will be a physical discontinuity in the centre of the ingot

Fig. 10.19 Hot rolling

256

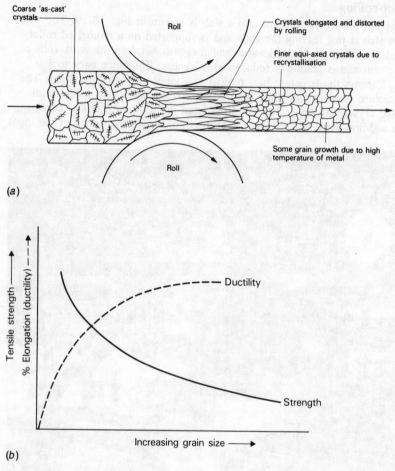

Coarse 'as-cast' crystals

Roll

Crystals elongated and distorted by rolling

Finer equi-axed crystals due to recrystallisation

Some grain growth due to high temperature of metal

Roll

(a)

Tensile strength ——→
% Elongation (ductility) ——→

Ductility

Strength

Increasing grain size ——→

(b)

Fig. 10.20 Effects of hot rolling (a) Hot-working a cast structure (b) Effect of grain size after hot-rolling on properties

called the *pipe*. These changes of structure and the associated discontinuities adversely affect the properties of the ingot. The top or neck of the ingot is removed and, with it, most of the pipe and any slag and impurities which have floated to the surface of the molten metal. The structure of the ingot is then rectified by the first hot-rolling process. The pipe and other internal discontinuities are pressure-welded by the heat of the ingot and the pressure of the rolls to make the metal homogeneous. The crystal structure is broken down and, since this is a hot-working process, recrystallisation occurs simultaneously to give a refined and uniform grain structure as shown in Fig. 10.20(a). The effect of hot-working on the strength and ductility of the metal is shown in Fig. 10.20(b).

Hot-extrusion

The principle of hot-extrusion is shown in Fig. 10.21. A hydraulic ram squeezes a billet of metal, which has been heated above its temperature of recrystallisation, through a die just like tooth-paste being squeezed out of the end of a tooth-paste tube. The hole in the die is shaped to produce the required section and long lengths of material are produced by this process. Materials which are commonly extruded are copper, brass alloys, aluminium, and aluminium alloys. After hot-extrusion the sections are often finished by cold drawing to improve the surface finish, dimensional accuracy, and stiffness.

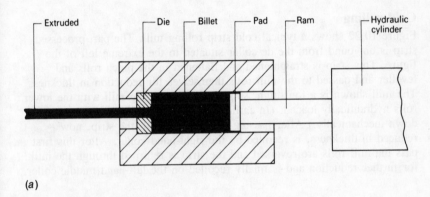

(a)

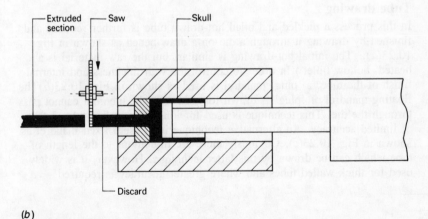

(b)

Fig. 10.21 Extrusion (a) Commencement of extrusion stroke (b) Completion of extrusion stroke

10.9 Some cold-working processes

Cold-working processes are essentially finishing processes. The forces required to produce quite modest reductions in cross-sectional area are very much higher than those for hot-working, so that the amount of reduction is kept to a minimum. However, as the finish and dimensional accuracy produced by cold-working is much superior to that produced by hot-working. Since cold-working results in some work-hardening of the metal, there is a corresponding improvement in its mechanical properties. The metal to be cold-worked is usually broken down by hot-working so that there is only a finishing allowance left. The oxide film (scale) on the surface of the hot-worked metal is removed by pickling the metal in acid, after which it is passed through a neutralising bath and oiled to prevent corrosion and passed to the cold-working process.

Cold-rolling

Figure 10.22 shows a typical cold strip rolling mill. The part-processed strip is unwound from the de-coiler situated to the extreme left of the figure. This strip is straightened and flattened in the pinch rolls and leveller and passed to the mill rolls themselves for reduction in thickness. The mill shown is a four-high, single-stand, reversing mill with the lower rolls hydraulically loaded. (In earlier mills the top rolls were screwed down mechanically.) After passing through the mill, the strip, now reduced in thickness, is recoiled on the right-hand coiler. After this first pass the mill rolls are reversed and the strip is returned through the mill for further reduction and is finally recoiled on the left-hand middle coiler.

Tube drawing

In this process a pickled and oiled hot-drawn tube is further reduced and finished by drawing it through a die on a draw-bench as shown in Fig. 10.23(a). (The initial hot-drawing is similar, but the raw material is a heated, hollow billet.) In order to control the wall thickness and internal finish of the tube, a 'plug' mandrel is used as shown in Fig. 10.23(b)The floating mandrel or 'plug' is drawn forward with the tube but cannot pass through the die. This technique is used for long, thin-walled tube, but is of limited accuracy. An alternative technique is to use a fixed mandrel as shown in Fig. 10.23(c). Obviously there are limitations to the length of tube which can be drawn by this latter technique. However, it is widely used for thick-walled tubes and where greater accuracy is required.

259

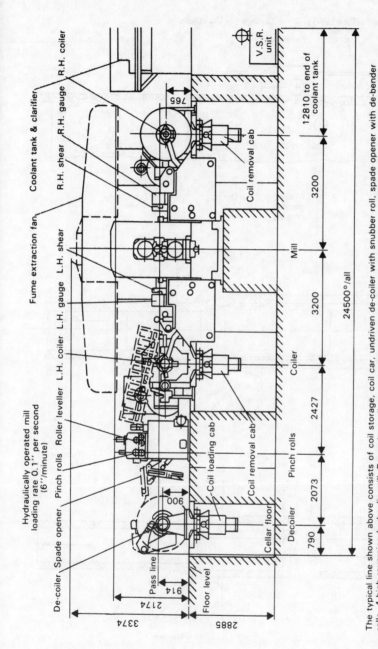

De-coiler, Spade opener, Pinch rolls, Roller leveller, L.H. coiler, L.H. gauge, L.H. shear

Fume extraction fan

R.H. shear, R.H. gauge, R.H. coiler

Coolant tank & clarifier

Hydraulically operated mill loading rate 0.1" per second (6"/minute)

V.S.R. unit

Coil removal cab

Mill

Coil loading cab

Coil removal cab

Decoiler   Pinch rolls   Coiler

Pass line

Floor level

Cellar floor

765

3200

3200

2427

2073

790

12810 to end of coolant tank

24500°/all

900

2174

3374

914

2885

The typical line shown above consists of coil storage, coil car, undriven de-coiler with snubber roll, spade opener with de-bender rolls, 4-high reversing hydraulic mill. Reversing coilers, coil car and storage station. Separate high and low pressure hydraulic packs provide the mill loading system and the operations of the ancillary equipment. A high capacity soluable oil system supplies strip lubrication and roll cooling.

Fig. 10.22 Cold rolling (courtesy Sir James Farmer Norton & Co. (International) Limited)

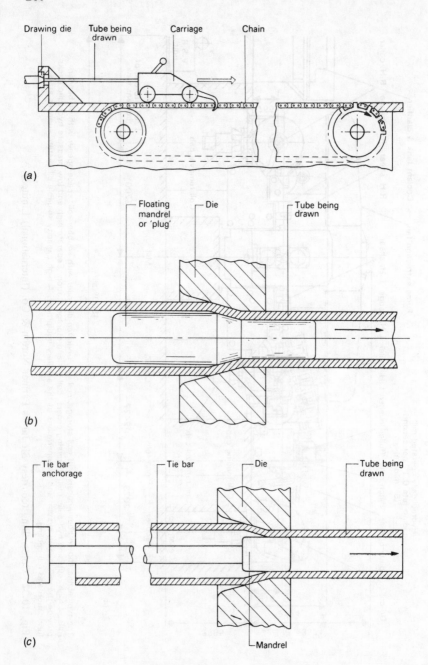

Fig. 10.23 Tube drawing (*a*) Simple draw bench (*b*) Tube drawing — use of a plug (*c*) Tube drawing over a mandrel

Wire drawing

This process is shown in Fig. 10.24(*a*). It is similar in principle to tube drawing, but since a longer length of material is involved, the wire is pulled through the die by a capstan or 'bull-block'. It may be coiled up on the capstan or passed to a separate coiler after taking only one or two turns round the capstan. Fine wire, as used for electrical conductors, is produced on multiple head machines as shown in Fig. 10.24(*b*). As the

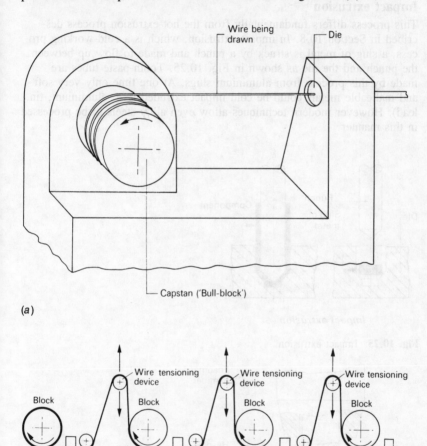

(*a*)

(*b*)

Fig. 10.24 Wire drawing (*a*) Single-die wire drawing (*b*) Multiple-die wire drawing

wire becomes thinner it becomes progressively longer. Thus each successive capstan has to run faster than the preceding one. The speed is controlled by the pull of the wire on the tension arm which is coupled to the motor speed control. If the tension on the arm increases, the capstan is slightly slowed but if the tension on the arm decreases, the capstan is speeded up.

Impact extrusion

This process differs fundamentally from the hot-extrusion process described in Section 10.8. In impact extrusion, which is a cold-working process, a slug of metal is struck by a punch and made to flow up between the punch and the die as shown in Fig. 10.25. Tooth-paste tubes are made by this process from aluminium slugs. At one time only very soft and malleable metals could be cold impact extruded (e.g. aluminium, tin, lead). However modern techniques allow even alloy steels to be processed in this manner.

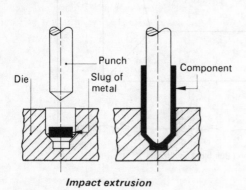

Impact extrusion

Fig. 10.25 Impact extrusion

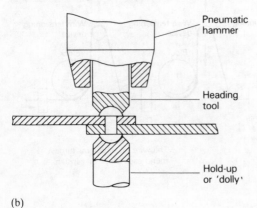

(b)

Fig. 10.26 Cold heading rivets

Rivet heading

This process is shown in Fig. 10.26. Here, the pre-formed head of the rivet is supported by a hold-up or 'dolly', whilst the opposite end of the rivet is headed by a pneumatic hammer. The shape of the rivet head being formed is controlled by the rivet 'snap' which is fitted to the hammer. In the example shown both ends of the rivet will be finished with a rounded or snap head.

The advantages and limitations of hot-working, cold-working and casting processes are compared in Tables 10.2, 10.3 and 10.4 respectively.

Table 10.2 Hot-working processes

Advantages	Limitations
1. Low cost	1. Poor surface finish – rough and scaly
2. Grain refinement from cast structure	2. Due to shrinkage on cooling the dimensional accuracy of hot-worked components is of a low order
3. Materials are left in the fully annealed condition and are suitable for cold-working (heading, bending, etc)	3. Due to distortion on cooling and to the processes involved, hot-working generally leads to geometrical inaccuracy
4. Scale gives some protection against corrosion during storage	4. Fully annealed condition of the material coupled with a relatively coarse grain leads to a poor finish when machined
5. Availability as sections (girders) and forgings as well as the more usual bars, rods, sheets and strip and butt welded tube	5. Low strength and rigidity for metal considered
	6. Damage to tooling from abrasive scale on metal surface

Table 10.3 Cold-working processes

Advantages	Limitations
1. Good surface finish	1. Higher cost than for hot-worked materials. It is only a finishing process for material previously hot-worked. Therefore, the processing cost is added to the hot-worked cost.
2. Relatively high dimensional accuracy	
3. Relatively high geometrical accuracy	2. Materials lack ductility due to work hardening and are less suitable for bending, etc
4. Work hardening caused during the cold-working processes: (*a*) increases strength and rigidity (*b*) improves the machining characteristics of the metal so that a good finish is more easily achieved	3. Clean surface is easily corroded
	4. Availability limited to rods and bars also sheets and strip, solid drawn tubes

Table 10.4 Casting processes (gravity, sand only)

Advantages	Limitations
1. Virtually no limit to the shape and complication of the component to be cast	1. Strength and ductility low, as structure is un-refined.
2. Virtually no limit to the size of the casting	2. Quality is uncertain, as local differences of structure and mechanical defects such as blow-holes cannot be controlled or corrected
3. Low cost, as no expensive machines and tools are required as in forging	3. Low accuracy due to shrinkage
4. Scrap metal can be reclaimed in the melting furnace. (Wrought and machined components have to be made from relatively expensive pre-processed materials)	4. Poor surface finish
	5. Component must be designed without sudden changes of section, so that molten metal flows easily and cooling cracks and warping will not occur
	6. Not all metals are suitable for casting. The best metals have a low shrinkage, a short freezing range and high fusibility (melt at relatively low temperatures), and have a high fluidity when molten

10.10 Moulding polymeric materials

Thermosetting plastics

These are usually moulded in dies fitted into hydraulic presses. A typical upstroke press is shown in Fig. 10.27. The platens are provided with electrical or steam heating elements so that the moulds fastened to them can be heated to the curing temperature of the thermosetting plastic being processed. Modern presses are automatic in operation and this ensures the constant moulding conditions at each stroke essential if mouldings of consistent quality are to be produced. The three factors which require to be pre-set to ensure correct polymerisation (curing) of the moulding powder are:

(a) the moulding pressure;
(b) the time for which the mould is closed and kept under pressure;
(c) the temperature of the mould.

Figure 10.28 shows a section through a simple flash-type mould. It is suitable for shallow components in which the thickness is not critical. To ensure complete filling of the mould a slight excess of moulding powder is placed in the mould cavity. The excess powder is forced out into the flash-gutter as the mould is closed and must be subsequently trimmed off

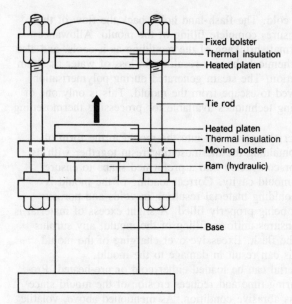

Fixed bolster
Thermal insulation
Heated platen

Tie rod

Heated platen
Thermal insulation
Moving bolster
Ram (hydraulic)

Base

Fig. 10.27 Moulding press

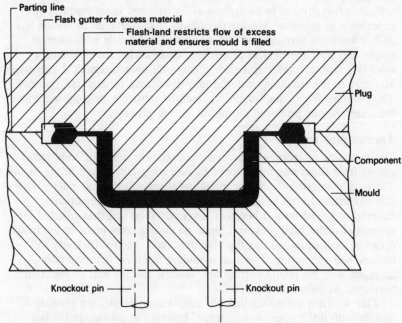

Parting line
Flash gutter for excess material
Flash-land restricts flow of excess material and ensures mould is filled

Plug

Component

Mould

Knockout pin
Knockout pin

Fig. 10.28 Flash mould

when the moulding is cold. The flash-land holds back the flow of the excess material and ensures complete filling of the mould. Allowance must be made for thermal shrinkage of the moulding as it cools, and also for shrinkage due to chemical changes resulting in loss of water as steam during the curing reaction. The steam generated during polymerisation (curing) must be allowed to escape from the mould. This is only one of many different moulding techniques available for processing thermosetting plastics.

The moulding material may be fed into the mould in the form of powder or granules containing the thermosetting resin together with its additives and fillers, or compacted into a preformed shape to ensure uniform filling of the mould cavity. Correct loading of the mould is critical, insufficient moulding material resulting in voids and porosity through the cavity not being properly filled. A slight excess of material is to be preferred as it ensures uniform filling of the mould, any surplus being allowed for in the flash. Excessive over-charging of the mould must be avoided as this can result in damage to the mould.

The moulding material can be loaded either cold or pre-heated. Pre-heating reduces the curing time and reduces erosion of the mould since the material is in a less abrasive condition. As mentioned above, volatile gases are released during curing and these must be allowed to escape through mould clearances and vents machined into the dies.

A release agent (lubricant) must be sprayed into the mould cavity to prevent sticking. The correct curing time and temperature, which is critical, generally has to be arrived at by trial and error based upon experience of previous, similar mouldings. An overcured moulding has a dull or blistered surface, and will be crazed and brittle with internal cracks and poor mechanical properties. Undercuring may produce a moulding with a correct appearance, but with poor mechanical properties. Moisture in the moulding material may also cause blisters and porosity. The moisture from damp moulding material must not be confused with the water vapour produced during the curing process.

Thermoplastics

The moulding of thermoplastic materials requires quite different techniques from those described above as no curing takes place in the mould. The most common process for moulding thermoplastic materials is *injection moulding*. In this process a measured amount of thermoplastic material is heated until it becomes a viscous fluid and then injected into the mould under high pressure. Since no curing takes place in the mould, it can be opened as soon as the moulding has cooled sufficiently to become self-supporting. Injection-moulding machines are generally arranged with the mould parting line vertical and the axis of injection horizontal as shown in Fig. 10.29.

Like so many processes that are simple in principle, the practice is fraught with difficulties. For instance, heating the plastic until it is a viscous fluid is not easy. If care is not taken the surface becomes

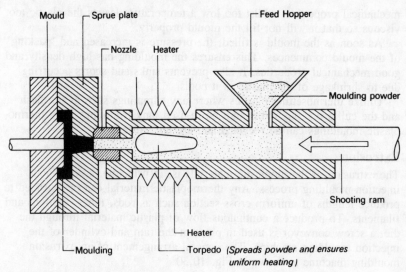

Fig. 10.29 Injection moulding

degraded by the heat before the inner mass of the plastic has reached moulding temperature. Again, the injected plastic material will not necessarily fill the mould cavity; there may be blisters, voids, sinks, shorts, and some cavities in a multi-impression mould may not receive any moulding material at all. Ejection of the completed moulding is also difficult if distortion is to be avoided and, since thermoplastic materials are adhesive, they will stick to the mould whenever the opportunity presents itself despite the use of a release agent. The following principal variables need to be controlled:

(*a*) the quantity of plastic material that is injected into the mould;
(*b*) the injection pressure;
(*c*) the injection speed;
(*d*) the temperature of the plastic moulding material;
(*e*) the temperature of the mould — too cold and the plastic may solidify before the mould is full; too hot and the moulding may be soft and lacking in rigidity when the mould is opened.
(*f*) the injection plunger forward-time. (The time the plastic material is maintained under pressure in the mould as it cools and becomes rigid. This reduces the effect of shrinkage and ensures the mould cavities are kept filled.)
(*g*) the mould-closed time;
(*h*) the mould clamping force;
(*i*) the mould-open time.

The temperature of the plastic is extremely critical as this controls its viscosity. Too high a temperature leads to degradation and poor mechanical properties, whilst too low a temperature leaves the plastic too viscous so that it will not fill the mould properly.

As soon as the mould is filled, the pressure is increased and 'packing' of the mould commences. This ensures the moulding has high density and good mechanical properties. It also prevents sinks and shorts occurring due to shrinkage of the plastic as it cools.

Unlike thermosetting plastics where any trimmings such as the flash and the cull are scrap and cannot be re-used, any trimmings from thermoplastic mouldings can be recycled and are not wasted.

Extrusion

The extrusion of thermoplastic materials is, in principle, a continuous injection-moulding process. Any thermoplastic material can be extruded to produce lengths of uniform cross-section such as rods, tubes, sections and filaments. To produce a continuous flow of plastic material through the die, a screw conveyor is used in place of the ram and cylinder of the injection moulding machine. The general arrangement of an extrusion moulding machine is shown in Fig. 10.30.

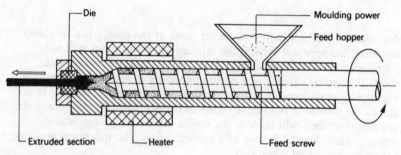

Fig. 10.30 Extrusion moulding

11 Materials testing (destructive)

11.1 Properties of materials

Although the properties of materials were introduced in Chapter 1, these properties are so heavily influenced by the composition, processing and heat treatment to which the material has been subjected, that further consideration of material properties and the testing of those properties has been delayed until this chapter. Now that the composition, processing and heat treatment of a range of metallic and non-metallic materials widely used by the engineer have been described, the problems and techniques associated with the testing of material properties can be more readily understood.

11.2 Tensile test

Strength is defined as the ability of a material to resist applied forces without yielding or fracturing. By convention strength usually denotes the resistance of a material to a tensile load applied axially to a specimen; this is the principle of the *tensile test*. Figure 11.1(a) shows a popular bench-mounted tensile testing machine, whilst Fig. 11.1(b) shows a more sophisticated machine suitable for industrial and research laboratories. This latter machine is capable of performing compression, shear and bending tests as well as tensile tests. Both these machines apply a carefully controlled tensile load to a standard specimen and measure the corresponding extension of that specimen.

Figure 11.2 shows some standard specimens and the direction of the applied load. These specimens are based upon British Standard BS 18. For the test results to be consistent for any given material, it is most important that the standard dimensions and profiles are adhered to. The shoulder radii are particularly critical and small variations, or the presence of tooling marks, can cause considerable differences in the test data obtained. Flat specimens are usually machined only on their edges so that the plate or sheet surface finish, and any structural deformation at the surface caused by the rolling process is taken into account in the test results.

Referring to Fig. 11.3, the *gauge length* (L_o) is the length over which the elongation of the specimen is measured. The *minimum parallel length* (L_c) is the minimum length over which the specimen must maintain a constant cross-sectional area before the test load is applied. The lengths L_o, L_c, L_1, and the cross-sectional area (a) are all specified in BS 18. Cylindrical test specimens are proportioned so that the gauge length L_o and the cross-sectional area a maintain a constant relationship. Hence such specimens are called *proportional test pieces*. The relationship is given by the expression:

$$L_o = 5.56\sqrt{a}$$

Since a = $0.25(\pi d^2)$
$\sqrt{a}$ = $0.886d$
Thus L_o = $5.56 \times 0.886d$
= $4.93d$
= $5d$ approx.

(a)

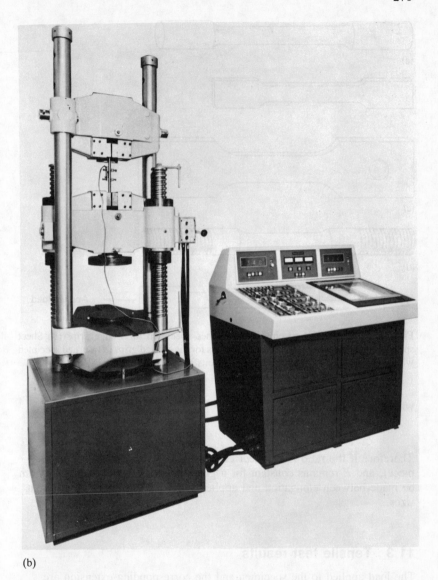

(b)

Fig. 11.1 Tensile testing machines (*a*) The Houndsfield tensometer (*b*) The Universal testing machine

Therefore a specimen 5 mm diameter will have a gauge length of 25 mm. The elongation obtained for a given force depends upon the length and area of cross-section of the specimen or component, since:

$$\text{Elongation} = \frac{\text{Force}}{E} \times \frac{L}{a}$$

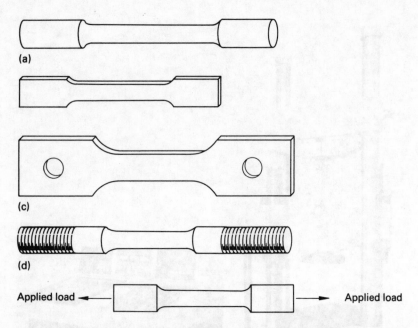

Fig. 11.2 Tensile test specimens (*a*) Turned specimen for wedge grips (*b*) Sheet specimen for wedge grips (*c*) Sheet specimens for pin-jointed grips (*d*) Turned specimen with screwed ends (For specimen dimensions and details see BS 18)

where L = length
a = cross-sectional area
E = elastic modulus

Therefore if the ratio L/a is kept constant (as it is in a proportional test piece), and E remains constant for a given material, then comparisons can be made between elongation and applied force for specimens of different sizes.

11.3 Tensile test results

The load applied to the specimen and the corresponding extension are plotted in the form of a graph as shown in Fig. 11.4.

(*a*) From *a* to *b* the extension is proportional to the applied load. Also, if the applied load is removed the specimen returns to its original length. Under these relatively lightly loaded conditions the material is showing *elastic* properties.

(*b*) From *b* to *c* it can be seen from the graph that the metal suddenly extends with no increase in load. If the load is removed at this point the

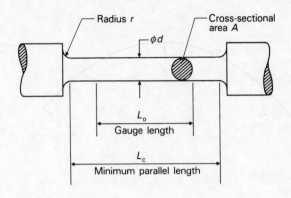

Fig. 11.3 Proportions for tensile test pieces (*a*) Cylindrical test piece (*b*) Flat test piece

metal will not spring back to its original length and it is said to have taken a *permanent set*. This is the *yield point* and the yield stress, which is the stress at the yield point, is the load at *b* divided by the original cross-section area of the specimen. Usually the designer works at 50 per cent of this figure to allow for a 'factor of safety' (see Section 13.1).

(*c*) From *c* to *d* extension is no longer proportional to the load and if the load is removed little or no spring back will occur. Under these relatively greater loads the material is showing *plastic* properties.

(*d*) The point *d* is referred to as the 'ultimate tensile strength' when referred to load/extension graphs or the 'ultimate tensile stress' (UTS)

274

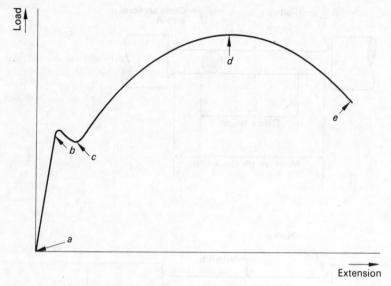

Fig. 11.4 Load/extension curve for low-carbon steel

when referred to stress/strain graphs. The ultimate tensile stress is calculated by dividing the load at d by the original cross sectional area of the specimen. Although a useful figure for comparing the relative strengths of materials, it has little practical value since engineering equipment is not usually operated so near to the breaking point.

(e) From d to e the specimen appears to be stretching under reduced load conditions. In fact the specimen is thinning out (*necking*) so that the 'load per unit area' or stress is actually increasing. The specimen finally work-hardens to such an extent that it breaks at e. In practice, values of load and extension are of limited use since they apply only to one particular size of specimen and it is more usual to plot the *stress/strain* curve. (An example of a stress/strain curve for a low-carbon steel is shown in Fig. 11.6.) Stress and strain are calculated as follows:

$$\text{Stress} = \frac{\text{load}}{\text{area of cross-section}}$$

$$\text{Strain} = \frac{\text{extension}}{\text{original length}}$$

11.4 Proof stress

Only very ductile materials such as fully annealed mild steel show a clearly defined yield point. The yield point will not even appear on bright drawn low-carbon steel which has become slightly work hardened during the drawing process. Under such circumstances the *proof stress* is used. proof stress is defined as the stress which produces a specified amount of plastic strain, such as 0.1 or 0.2 per cent. Figure 11.5 shows a typical stress/strain curve for a material of relatively low ductility, such as hardened and tempered medium carbon steel. If a point such as *C* is taken, the corresponding strain is given by *D* and this consists of a combination of plastic and elastic components. If the stress is now gradually reduced (by reducing the load on the specimen), the strain is also reduced and the stress/strain relationship during this reduction in stress is represented by the line *CB*. During the reduction in stress the elastic deformation is recovered so that the line *CB* is straight and parallel to the initial stages of the loading curve for the material, that is, the part of the loading curve where the material is showing elastic properties.

In the example shown, the stress at *C* has produced a plastic strain of 0.2 per cent as represented by *AB*. Thus the stress at *C* is referred to as *0.2 per cent Proof Stress, AB* being the plastic deformation and *BD* being the elastic deformation when the specimen is stressed to the point *C*. The

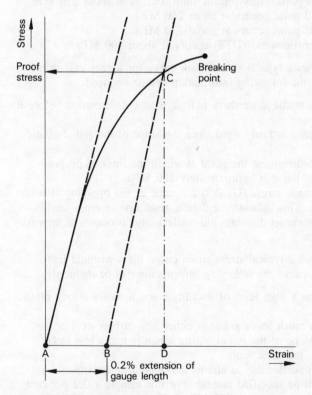

Fig. 11.5 Proof stress

material will have fulfilled its specification if, after the proof stress has been applied for 15 seconds and removed, the permanent set of the specimen is not greater than the specified percentage of the gauge length which, in this example, is 0.2 per cent.

11.5   The interpretation of tensile test results

The interpretation of tensile test data requires skill born out of experience, since many factors can affect the test results, for instance the temperature at which the test is carried out, since the tensile modulus and tensile strength decrease as the temperature rises for most metals and plastics, whereas the ductility increases as the temperature rises. The test results are also influenced by the rate at which the specimen is strained.

Figure 11.6(a) shows a typical stress/strain curve for an annealed mild steel. From such a curve the following information can be deduced.

(a) The material is ductile since there is a long plastic range.
(b) The material is fairly rigid since the slope of the initial elastic range is steep.
(c) The limit of proportionality (elastic limit) occurs at about 230 MPa.
(d) The upper yield point occurs at about 260 MPa.
(e) The lower yield point occurs at about 230 MPa.
(f) The ultimate tensile stress (UTS) occurs at about 400 MPa.

Figure 11.6(b) shows a typical stress/strain curve for a grey cast iron. From such a curve the following information can be deduced.

(a) The material is brittle since there is little plastic deformation before it fractures.
(b) Again the material is fairly rigid since the slope of the initial elastic range is steep.
(c) It is difficult to determine the point at which the limit of proportionality occurs, but it is approximately 200 MPa.
(d) The ultimate tensile stress (UTS) is the same as the breaking stress for this sample. This indicates negligible reduction in cross-section (necking) and minimal ductility and malleability. It occurs at approximately 250 MPa.

Figure 11.6(c) shows a typical stress/strain curve for a wrought light alloy. From such a curve the following information can be deduced.

(a) The material has a high level of ductility since it shows a long plastic range.
(b) The material is much less rigid than either low-carbon steel or cast iron since the slope of the initial plastic range is much less steep when plotted to the same scale.
(c) The limit of proportionality is almost impossible to determine, so *proof stress* will be specified instead. For this sample a 0.2 per cent proof stress is approximately 500 MPa (AB).

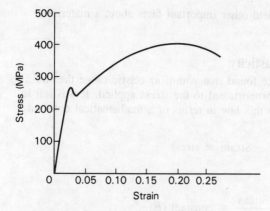

(a) Stress/strain curve for annealed mild steel

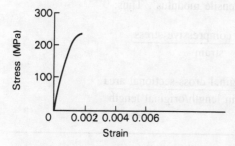

(b) Stress/strain curve for grey cast iron

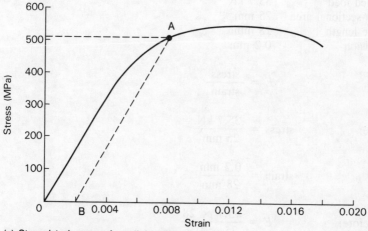

(c) Stress/strain curve for a light alloy

Fig. 11.6 Typical stress/strain curves

The tensile test can also yield other important facts about a material under test.

Young's modulus of elasticity

The physicist Robert Hooke found that within its elastic range the strain produced in a material is proportional to the stress applied. It was left to Thomas Young to quantify this law in terms of a mathematical constant for any given material.

$$\text{Strain} \propto \text{stress}$$

Therefore:

$$\frac{\text{Stress}}{\text{Strain}} = \text{constant } (E)$$

This constant term (E) is variously known as 'Young's modulus', the 'modulus of elasticity', or the 'tensile modulus'. Thus:

$$E = \frac{\text{tensile or compressive stress}}{\text{strain}}$$

$$= \frac{\text{force/original cross-sectional area}}{\text{change in length/original length}}$$

Example 11.1 Calculate the modulus of elasticity for a material which produced the following data when undergoing a tensile test.

Applied load	35.7 kN
Cross-sectional area	25 mm^2
Gauge length	28 mm
Extension	0.2 mm

$$E = \frac{\text{stress}}{\text{strain}}$$

where:

$$\text{stress} = \frac{35.7 \text{ kN}}{25 \text{ mm}^2}$$

$$\text{strain} = \frac{0.2 \text{ mm}}{28 \text{ mm}}$$

Therefore:

$$E = \frac{35.7 \times 28}{25 \times 0.2}$$

$$= 199.92 \text{ kN/mm}^2$$
$$\underline{= 200 \text{ GPa (approx.)}}$$

This would be a typical value for a low-carbon steel.

It has already been stated (Section 1.2) that malleability and ductility are special cases of the general property of plasticity. It was stated that *malleability* refers to the extent to which a material can undergo deformation in compression before failure occurs, and that *ductility* refers to the extent to which a material can undergo deformation in tension before failure occurs. All ductile materials are malleable, but all malleable materials are not ductile since they may lack the strength to withstand tensile loading.

Therefore ductility is usually expressed, for practical purposes, as the percentage elongation in gauge length of a standard test piece at the point of fracture when subjected to a tensile test to destruction.

$$\text{Elongation \%} = \frac{\text{increase in length} \times 100}{\text{original length}}$$

The increase in length is determined by fitting the pieces of the fractured specimen together carefully and measuring the length at failure.

Increase in length (elongation) = length at failure − original length

Figure 11.7 shows a specimen of a soft, ductile material before and after testing. It can be seen that the specimen does not reduce in cross-sectional area uniformly, but that severe local necking occurs prior to fracture. Since most of the plastic deformation and, therefore, most of the elongation occurs in the necked region, doubling the gauge length does not result in double the elongation when calculated as a percentge of gauge length. Therefore it is important to use a standard gauge length if

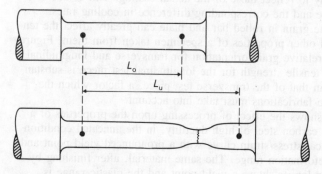

Fig. 11.7 Elongation

comparability between results is to be achieved. Elongation is calculated as:

$$\text{Elongation} \ \% = \frac{(L_u - L_o)}{L_o} \times 100$$

Example 11.2 Calculate the percentage elongation for a 70/30 brass alloy if the original gauge length (L_o) is 56 mm and the length at fracture is (L_u) is 95.2 mm.

$$\text{Elongation} \ \% = \frac{L_u - L_o}{L_o} \times 100$$

$$= \frac{(95.2 - 56)}{56} \times 100$$

$$= \underline{70\%}$$

Table 11.1 lists some typical test results for a range of metallic materials.

11.6 The effect of grain size and structure on tensile testing

The test piece should be chosen so that it reflects as closely as possible the component and the material from which the component is produced. This is relatively easy for components produced from bar stock, but not so easy for components produced from forgings as the grain flow will be influenced by the contour of the component and will not be uniform. Castings also present problems since the properties of a specially cast test piece are unlikely to reflect those of the actual casting. This is due to the difference in size and the corresponding difference in cooling rates.

The lay of the grain in rolled bar and plate can greatly affect the tensile strength and other properties of a specimen taken from them. Figure 11.8 shows the relative grain orientation for transverse and longitudinal test pieces. The tensile strength for the longitudinal test piece is substantially greater than that of the transverse test piece, a factor which the designer of large fabrications must take into account.

Figure 11.9 shows the effect of processing upon the properties of a material. A low carbon steel of high ductility, in the annealed condition shows the classical stress/strain curve with a pronounced yield point and a long plastic deformation range. The same material, after finishing by cold-drawing, no longer shows a yield point and the plastic range is noticeably reduced.

Table 11.1 Typical tensile test results for metallic materials

Material	Condition	Tensile strength			Elongation
		Yield point (MPa)	0.1% PS* (MPa)	UTS† (MPa)	(%)
Aluminium bronze	Annealed	—	125	385	70
	Hard	—	590	775	4
Cartridge brass (70/30)	Annealed	—	77	325	70
	Hard	—	510	695	5
Commercial brass (63/37)	Annealed	—	95	340	55
	Hard	—	540	725	4
Muntz brass (60/40)	Extruded	—	110	370	40
Cupro-nickel (256 Ni)	Annealed	—	—	355	45
	Hard	—	—	600	5
Duralumin	Age hardened	—	280	400	10
Grey cast iron	—	—	—	150/250	Negligible
Low tin bronze	Annealed	—	110	340	65
	Hard	—	620	740	5
Magnesium alloy	Wrought	—	155	280	10
Medium carbon steel	Quench hardened and tempered at 550–650 °C	550	—	750	14
Low carbon steel	Annealed	250/350	—	450/525	20/25
Stainless steel (ferritic)	Annealed	340	—	510	31
Stainless steel (Austentic 18/8)	Annealed	278	—	618	50
	Cold-rolled	803	—	896	30

*PS = Proof stress †UTS = Ultimate tensile strength

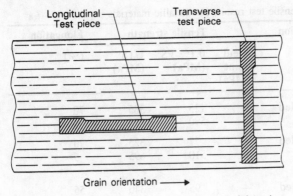

Longitudinal Test piece — Transverse test piece

Grain orientation ——➤

Fig. 11.8 Effect of grain orientation on material testing

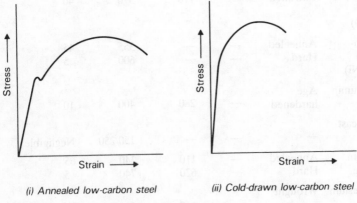

(i) Annealed low-carbon steel   (ii) Cold-drawn low-carbon steel

Fig. 11.9 Effect of processing on the properties of a low-carbon steel

Figure 11.10 shows the effect of heat treatment upon the properties of a medium carbon steel. In this example the results have been obtained by quench hardening a batch of identical specimens and then tempering them at different temperatures.

Figure 11.11 shows the effect of heat treatment upon the properties of a work-hardened metallic material. Stress relief (recovery) has very little effect upon the tensile strength and elongation (ductility) until the recrystallisation (annealing) temperature is reached. The metal initially shows the high tensile strength and lack of ductility associated with a severely distorted grain structure. After stress relief the tensile strength rises and the ductility falls until the recrystallisation temperature is reached. During the recrystallisation range there is a marked change in properties. The tensile strength is rapidly reduced and the ductility, in terms of elongation percentage, rapidly increases.

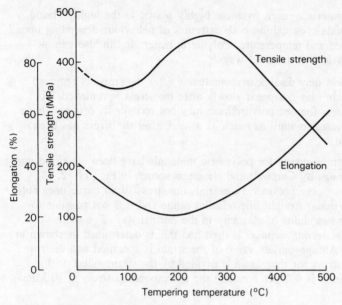

Fig. 11.10 Effect of tempering on tensile test results

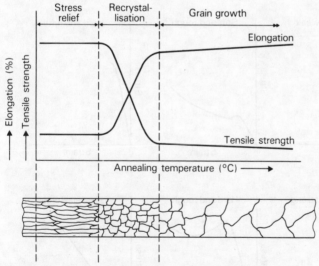

Fig. 11.11 Effect of temperature on cold-worked materials

11.7 Tensile testing polymeric materials

The tensile test can also be used to determine the properties of polymeric materials. However, some care is required in interpreting the results since

284

polymeric materials range from the highly plastic to the highly elastic. Some materials even exhibit both extremes of behaviour depending upon the strain rate and temperature. Polymeric materials can also exhibit elastic strain in two different ways.

(a) The strain may disappear immediately after the stress is removed.
(b) The strain may disappear slowly after the stress is removed. For example, a foamed polyurethane may not recover its original shape and dimensions until as much as a *week* after the stress has been removed.

The stress/strain curves for polymeric materials have been classified into six main groups by Carswell and Nason as shown in Fig. 11.12. However, for some polymeric materials the stress/strain curve does not indicate the initial straight line (elastic) range and it is not possible to determine the modulus of elasticity in the normal way. For such materials, the *secant modulus* is used and this is determined as shown in Fig. 11.13. An appropriate value of the strain is specified and the corresponding stress for that strain is divided by the strain value. In Fig. 11.13 the strain is 0.2 per cent and the corresponding stress is 30 MPa. Hence:

$$\text{Secant modulus} = \frac{30 \times 100}{0.2} = 15 \text{ GPa}$$

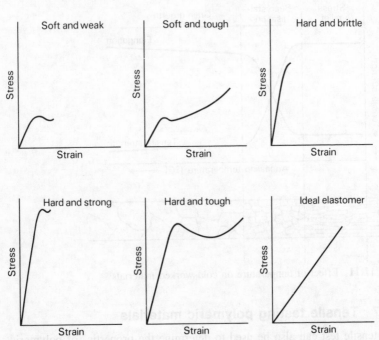

Fig. 11.12 Typical stress/strain curves for polymers (after Carswell & Nason)

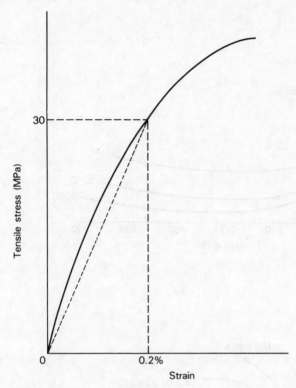

Fig. 11.13 Secant modulus

The test results obtained for polymeric materials are much more influenced by the test conditions than are the corresponding tests for metallic materials. Figure 11.14(a) shows how the temperature can affect the test results obtained. For most rigid polymers the percentage elongation increases as the temperature increases, but the strength is much lower. At lower temperatures (at or below the T_g) the material becomes brittle and fails with negligible elongation. For some rubbers the elongation can actually increase at low temperatures within limits.

The strain rate can also have a significant effect on the results obtained from specimens of a given material. Figure 11.14(b) shows how the strength of the material appears to increase when straining occurs at a lower rate (over a longer time).

The stress/strain behaviour of a typical elastomer (polyisobutylene rubber) is shown in Fig. 11.15. It can be seen that in elastomers strain is not proportional to stress, but that they show the 'S'-shaped relationship typical of rubbers. This is, of course, completely unlike the linear stress/strain relationship of thermosets and metals when stressed within their elastic limits.

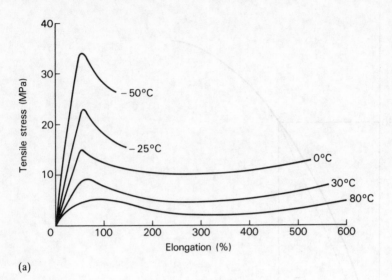

(a)

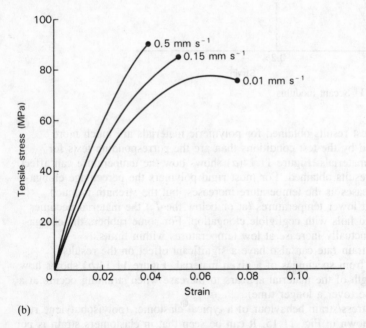

(b)

Fig. 11.14 Effects of test conditions on polymers (a) Effect of temperature on tensile strengths of low-density polythene (b) Effect of strain rate on tensile strength of Perspex

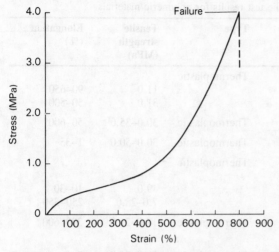

Fig. 11.15 Stress/strain curve for polyisobutylene rubber

Creep is also an important factor when testing polymeric materials. It is defined as the gradual extension of a material over a period of time when subjected to a constant applied load, and this will be considered in detail in Section 13.2. Table 11.2 lists some typical tensile test results for a range of polymeric materials.

11.8 Impact testing

The tensile test does not tell the whole story. Figure 11.16 shows how a piece of high-carbon steel rod will bend when in the annealed condition, yet snap easily in the quench-hardened condition despite the fact that in the latter condition it will show a much higher value of tensile strength. Impact tests consist of striking a suitable specimen with a controlled blow and measuring the energy absorbed in bending or breaking the specimen. The energy value indicates the toughness of the material under test. Figure 11.17 shows a typical impact testing machine. This machine has a hammer which is suspended like a pendulum, a vice for holding the specimen in the correct position relative to the hammer and a dial for indicating the energy absorbed in carrying out the test in joules (J). If there is maximum overswing, as there would be if no specimen was placed in the vice, then zero energy absorption is indicated. If the hammer is stopped by the specimen with no overswing, then maximum energy absorption is indicated. Intermediate readings are the *impact values* (J) of the materials being tested (their toughness or lack of brittleness). There are two standard tests currently in use.

Table 11.2 Typical tensile test results for polymeric materials

Material	Type	Tensile strength (MPa)	Elongation (%)
Polythene (low density) (high density)	Thermoplastic	11.0 31.0	90–650 50–500
Polypropylene	Thermoplastic	30.0–35.0	50–600
Polystyrene	Thermoplastic	30.0–50.0	1–35
Polyvinyl chloride (PVC) (rigid) (flexible)	Thermoplastic	49.0 7.0–25.0	10–30 250–350
Polytetrafluoroethylene (PTFE)	Thermoplastic	17.0–25.0	200–600
Polymethyl methacrylate (Perspex)	Thermoplastic	50.0–70.0	3–8
Polyamide (nylon '66')	Thermoplastic	50.0–87.5	60–300
Polycarbonate	Thermoplastic	60.0–70.0	60–100
Cellulose acetate	Thermoplastic	24.0–65.0	5–55
Phenol formaldehyde	Thermosetting plastic	35–55 (wood floor filler)	1
Urea formaldehyde	Thermosetting plastic	50–75 (cellulose filler)	1
Melamine formaldehyde	Thermosetting plastic	55–80 (cellulose filler)	0.7
Epoxy resin rigid and unfilled	Thermosetting plastic	35–80	5–10

The Izod test

In this test a 10 mm square, notched specimen is used. The striker of the pendulum hits the specimen with a kinetic energy of 162.72 J at a velocity of 3.8 m/s. Figure 11.18 shows details of the specimen and the manner in which it is supported.

The Charpy test

In the Izod test the specimen is supported as a cantilever, but in the Charpy test it is supported as a beam. It is struck with a kinetic energy

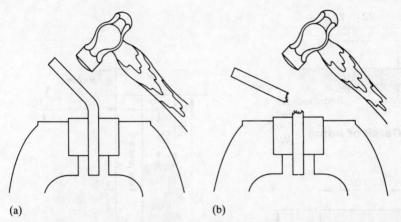

(a)   (b)

Fig. 11.16 Impact loading (*a*) A piece of high-carbon steel rod (1.0%) in the annealed (soft) condition will bend when struck with a hammer. UTS 925 MPa (*b*) The same piece of high-carbon steel rod, as in (*a*), after hardening and lightly tempering will fracture when hit with a hammer despite its UTS having increased to 1285 MPa

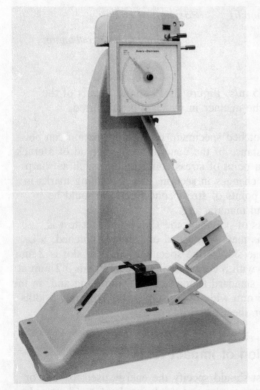

Fig. 11.17 Typical impact testing machine

290

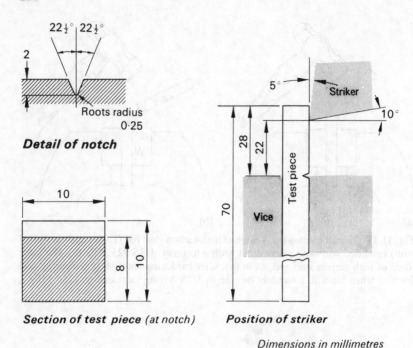

Detail of notch

Section of test piece (at notch)   **Position of striker**

Dimensions in millimetres

Fig. 11.18 Izod test

of 298.3 J at a velocity of 5 m/s. Figure 11.19 shows details of the
Charpy test specimen and the manner in which it is supported.

Since both tests use a notched specimen, useful information can be
obtained regarding the resistance of the material to the spread of a crack
which may originate from a point of stress concentration such as sharp
corners, undercuts, sudden changes in section, and machining marks in
stressed components. Such points of stress concentration should be
eliminated during design and manufacture.

For testing the toughness of polymers the *Charpy* impact test is
usually used. The specimens may be notched or plain. If notched, a U-
shaped notch is milled in one side of the specimen, and the slot is 2 mm
wide, 3 mm or 5 mm deep with a corner radius not less than 0.2 mm at
the bottom of the slot. The standard testpiece is 120 mm long and, in the
case of moulded plastic, 15 mm wide by 10 mm thick. Different widths
and thicknesses are used for sheet plastic.

11.9 The interpretation of impact tests

The results of an impact test should specify the energy used to bend or
break the specimen and the particular test used, i.e. Izod or Charpy. In

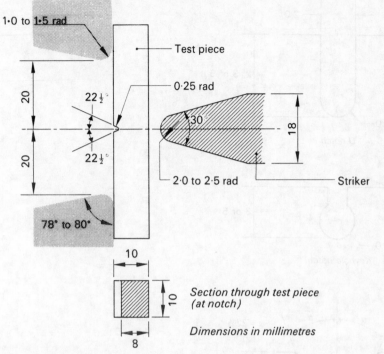

Fig. 11.19 Charpy test

the case of the Charpy test it is also necessary to specify the type of notch used as this test allows for three types of notch, as shown in Fig. 11.20. A visual examination of the fractured surface after the test also provides useful information.

(a) *Brittle Metals.* A clean break with little deformation and little reduction in cross-sectional area at the point of fracture. The fractured surfaces will show a granular structure.

(b) *Ductile metals.* The fracture will be rough and fibrous. In very ductile materials the fracture will not be complete, the specimen bending over and only showing slight tearing from the notch. There will also be some reduction in cross-sectional area at the point of fracture or bending.

(c) *Brittle Polymers.* A clean break showing smooth, glassy, fractured surfaces with some splintering.

(d) *Ductile Polymers.* No distinctive appearance to the fracture except for a considerable reduction in cross-sectional area and some tearing of the notch. Very ductile polymers may simply bend with some tearing at the notch but no complete fracture will occur.

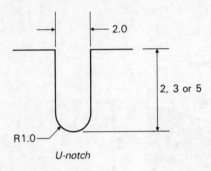

2.0

2, 3 or 5

R1.0

U-notch

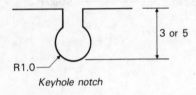

3 or 5

R1.0

Keyhole notch

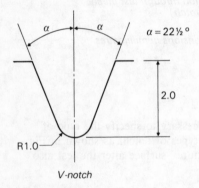

α α $\alpha = 22\frac{1}{2}°$

2.0

R1.0

V-notch

Dimensions in millimetres

Fig. 11.20 Standard Charpy notches

The temperature of the specimen at the time of making the test also has an important influence on the test results. Figure 11.21 shows the embrittlement of low carbon steels at refrigerated temperatures and hence their unsuitability for use in refrigeration plant and space vehicles. Some typical impact test results for a range of metallic materials and a range of polymers are given in Tables 11.3 and 11.4. Whilst the energy absorbed on impact for metallic specimens is usually stated in joules (J), the impact energy absorbed by polymer specimens is often divided by the cross-sectional area of the specimen for un-notched specimens, or the area behind the notch for notched specimens.

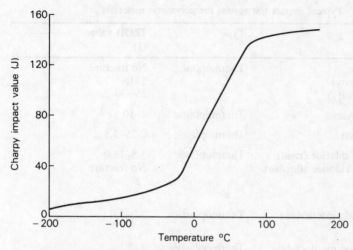

Fig. 11.21 Effect of test temperature on toughness

Table 11.3 Typical impact test results for metallic materials

Material	Condition	Charpy V (J)
Aluminium bronze	Annealed	100
	Hard	25
Cartridge brass (70/30)	Annealed	88
	Hard	20
Commercial brass (63/37)	Annealed	65
	Hard	12
Muntz brass (60/40)	Extruded	6
Cupro-nickel (25% Ni)	Annealed	157
	Hard	35
Duralumin	Age hardened	80
Grey cast iron	—	2.5
Magnesium alloy	Wrought	12
Medium carbon steel	Quench hardened and tempered at 550–650 °C	65
Low carbon steel	As hot rolled	50
Stainless steel (ferritic)	Annealed	165
Stainless steel (austenitic)	Annealed	217

Table 11.4 Typical impact test results for polymeric materials

Material	Type	IZOD value (J)
Polythene (low density) (high density)	Thermoplastic	No fracture 2–18 25–30
Polypropylene	Thermoplastic	1–10
Polystyrene	Thermoplastic	0.25–2.5
Polyvinyl chloride (rigid) Polyvinyl chloride (flexible)	Thermoplastic	1.5–18.0 No fracture
Polytetrafluoroethylene (PTFE)	Thermoplastic	3.0–5.0
Polymethyl methacrylate (perspex)	Thermoplastic	3–8
Polyamide (nylon '66')	Thermoplastic	1.5–15.0
Polycarbonate	Thermoplastic	10–20
Cellulose acetate	Thermoplastic	5–55
Phenol formaldehyde	Thermosetting plastic	0.3–1.0 wood floor filler
Urea formaldehyde	Thermosetting plastic	0.3–0.5 cellulose filler
Melamine formaldehyde	Thermosetting plastic	0.2–0.5 cellulose filler
Epoxy resin – rigid and unfilled	Thermosetting plastic	0.5–1.5

11.10 The effect of processing on toughness

Impact tests are frequently used to determine the effectiveness of annealing temperatures on the grain structure and impact strength of cold-worked, ductile metals. In the case of cold-worked low-carbon steel, the impact strength is quite low, initially, as the heavily deformed grain structure will be relatively brittle and lacking in ductility, particularly if the limit of cold-working has been approached. Annealing at low temperatures has little effect as it only promotes recovery of the crystal lattice on the atomic scale and does not result in recrystallisation. In fact during recovery there may even be a slight reduction in the impact strength.

However, at about 550°C to 650°C recrystallisation of low-carbon steels occurs with only slight grain growth. Annealing in this temperature range results in the impact strength increasing dramatically as shown in Fig. 11.22, and the appearance of the fracture changes from that of a brittle material to that of a ductile material. Annealing at higher temperatures or prolonged soaking at the lower annealing temperature results in grain growth and a corresponding fall in impact strength.

The effect of tempering on the impact value of a quench-hardened high-carbon steel is shown in Fig. 11.23. Initially, only stress relief occurs but as the tempering temperature increases, the toughness also increases which is why cutting tools are tempered. Tempering modifies the extremely hard and brittle martensitic structure of quench-hardened plain-carbon steels and causes a considerable increase in toughness with very little loss of hardness.

11.11 Hardness testing

Hardness has already been defined in Section 1.2 as the resistance of a material to indentation or abrasion by another hard body. It is by indentation that most hardness tests are performed. A hard indenter is pressed into the specimen by a standard load, and the magnitude of the indentation (either area or depth) is taken as a measure of hardness.

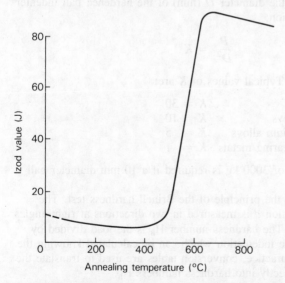

Fig. 11.22 Effect of annealing on the toughness of a low-carbon steel

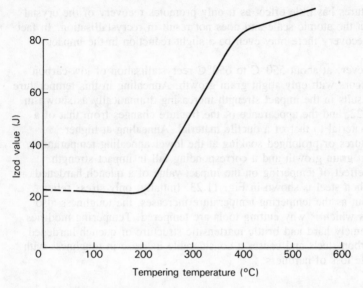

Fig. 11.23 Effect of tempering on the toughness of a quench-hardened high carbon steel

The Brinell hardness test

In this test, hardness is measured by pressing a hard steel ball into the surface of the test piece, using a known load. It is important to choose the combination of load and ball size carefully so that the indentation is free from distortion and suitable for measurement. The relationship between load P (kg) and the diameter D (mm) of the hardened ball indenter is given by the expression:

$$\frac{P}{D^2} = K$$

where K is a constant. Typical values of K are:

Ferrous metals	$K = 30$
Copper and copper alloys	$K = 10$
Aluminium and aluminium alloys	$K = 5$
Lead, tin, and white bearing metals	$K = 1$

Thus, for steel, a load of 3000 kg is required if a 10 mm diameter ball indenter is used.

Figure 11.24 shows the principle of the Brinell hardness test. The diameter of the indentation d is measured in two directions at right-angles and the average taken. The hardness number H_B is the load divided by the spherical area of the indentation which can be calculated knowing the values of d and D. In practice, conversion tables are used to translate the value of diameter d directly into hardness numbers H_B.

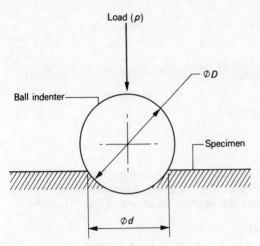

Fig. 11.24 Brinell hardness (principle)

To ensure consistent results the following precautions should be observed:

(a) the thickness of the specimen should be at least seven times the depth of the indentation to allow unrestricted plastic flow below the indenter;
(b) the edge of the indentation should be at least three times the diameter of the indentation from the edge of the test piece;
(c) the test is unsuitable for materials whose hardness exceeds 500 H_B, as the ball indenter tends to flatten.

Data other than hardness can be obtained from this test as follows.

Machinability

With high-speed steel cutting tools, the hardness of the stock being cut should not exceed $H_B = 350$ for a reasonable tool life. Again, materials with a hardness less than $H_B = 100$ will tend to tear and leave a poor surface finish.

Tensile strength

There is a definite relationship between strength and hardness, and the ultimate tensile stress (UTS) of a component can be approximated as follows:

$$\text{UTS (MPa)} = H_B \times 3.54 \text{ (for annealed plain-carbon steels);}$$
$$= H_B \times 3.25 \text{ (for quench-hardened and tempered}$$
$$\text{plain-carbon steels);}$$
$$= H_B \times 5.6 \text{ (for ductile brass alloys);}$$
$$= H_B \times 4.2 \text{ (for wrought aluminium alloys).}$$

(a) Piling up	(b) Sinking

Fig. 11.25 Work-hardening capacity

Work-hardening capacity

Materials which will cold-work without work-hardening unduly will pile up round the indenter as shown in Fig. 11.25(*a*). Materials which work-harden readily will sink around the indenter as shown in Fig. 11.25(*b*).

The Vickers hardness test

This test is preferable to the Brinell test where hard materials are concerned, as it uses a diamond indenter. (Diamond is the hardest material known — approximately 6000 H_B.) The diamond indenter is in the form of a square-based pyramid with an angle of 136° between opposite faces. Since only one type of indenter is used the load has to be varied for different hardness ranges. Standard loads are 5, 10, 20, 30, 50 and 100 kg. It is necessary to state the load when specifying a Vickers hardness number. For example if the hardness number is found to be 200 when using a 50 kg load, then the hardness number is written $H_D(50) = 200$. Figure 11.26(*a*) shows a universal hardness testing machine suitable for performing both Brinell and Vickers hardness tests, whilst Fig. 11.26(*b*) shows the measuring screen for determining the distance across the corners of the indentation. The screen can be rotated so that two readings at right-angles can be taken and the average is used to determine the hardness number (H_D). This is calculated by dividing the load by the projected area of the indentation:

$$H_D = P/D^2,$$

where D = the average diagonal (mm), P = load (kg).

The Rockwell hardness test

Although not so reliable as the Brinell and Vickers hardness tests for laboratory purposes, the Rockwell test is widely used in industry as it is quick, simple, and direct-reading. Figure 11.27(*a*) shows a typical Rockwell hardness testing machine, whilst Fig. 11.27(*b*) shows a typical hardness indicating scale. Universal electronic hardness testing machines are now widely used which, at the turn of a switch, can provide either Brinell, Vickers, or Rockwell tests and which show the hardness number as a digital readout automatically. They also give a 'hard copy' print-out of the test result together with the test conditions and date. However, the mechanical testing machines described in this chapter are still widely used and will be for some time to come.

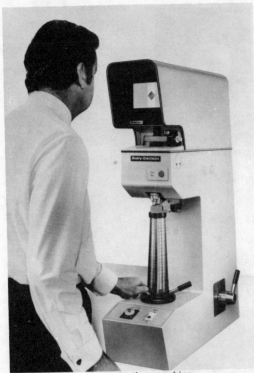

(a) Universal hardness testing machine

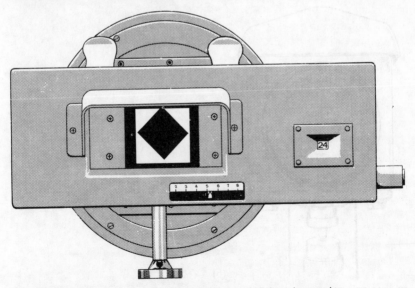

(b) Measuring screen showing magnified image of Vickers impression

Fig. 11.26 Hardness testing

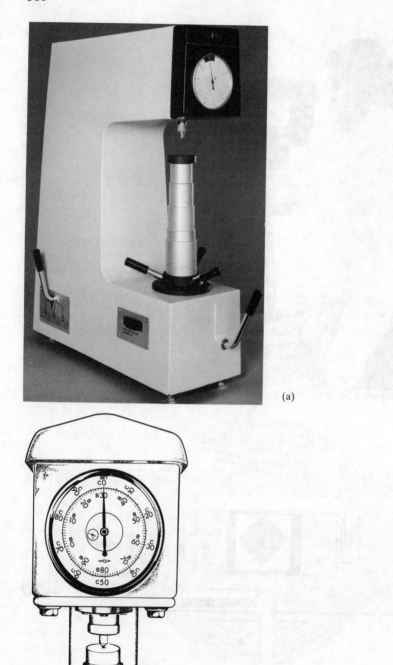

Fig. 11.27 Rockwell hardness test (*a*) Rockwell direct reading testing machine (*b*) Direct reading hardness scale

In principle the Rockwell hardness test compares the difference in *depth* of penetration of the indenter when using forces of two different values, that is, a minor force is first applied (to take up the backlash and pierce the skin of the component) and the scales are set to read zero. Then a major force is applied over and above the minor force and the increased depth of penetration is shown on the scales of the machine as a direct reading of hardness without the need for calculation or conversion tables.

The indenters most commonly used are a 1.6 mm diameter hard steel ball and a diamond cone with an apex angle of 120°. The minor force in each instance is 98 N. Table 11.5 gives the combinations of type of indenter and additional (major) force for the range of Rockwell scales, together with typical applications. The B and C scales are the most widely used in engineering.

The standard Rockwell test cannot be used for very thin sheet and foils, and for these the *Rockwell Superficial Hardness Test* is used. The minor force is reduced from 98 N to 29.4 N and the major force is also reduced. Typical values are listed in Table 11.6.

Shore scleroscope

In the tests previously described, the test piece must be small enough to mount in the testing machine, and hardness is measured as a function of indentation. However the scleroscope, shown in Fig. 11.28, works on a different principle, and hardness is measured as a function of resilience. Further, since the scleroscope can be carried to the work piece, it is useful for testing large surfaces such as the slideways on machine tools. A diamond-tipped hammer of mass 2.5 g drops through a height of 250 mm. The height of the first rebound indicates the hardness on a 140-division scale.

11.12   The effect of processing on hardness

All metals work-harden to some extent when cold-worked. Figure 11.29 shows the relationship between the Vickers hardness number (H_D) and the percentage reduction in thickness for rolled strip. The metals become harder and more brittle as the amount of cold-working increases until a point is reached where the metal is so hard and brittle that cold-working cannot be continued. Aluminium reaches this state when a 60 per cent reduction in strip thickness is achieved in one pass through the rolls of a rolling mill. In this condition the material is said to be fully work-hardened. The degree of work-hardening or 'temper' of strip and sheet material is arbitrarily stated as soft (fully annealed), 1/4 hard, 1/2 hard, 3/4 hard, and hard (fully work-hardened).

The effect of heating a work-hardened material such as α brass is shown in Fig. 11.30. Once again very little effect occurs until the temperature of recrystallisation is reached. At this temperature there is a rapid fall off in hardness, after which the decline in hardness becomes

Table 11.5 Rockwell hardness test conditions

Scale	Indenter	Additional force (kN)	Applications
A	Diamond cone	0.59	Steel sheet; shallow case-hardened components
B	Ball, ∅ 1.588 mm	0.98	Copper alloys; aluminium alloys, and annealed low carbon steels
C	Diamond cone	1.47	Most widely used range: hardened steels; cast irons; deep case-hardened components
D	Diamond cone	0.98	Thin but hard steel – medium depth case-hardened compounds
E	Ball, ∅ 3.175 mm	0.98	Cast iron, aluminium alloys; magnesium alloys, bearing metals
F	Ball, ∅ 1.588 mm	0.59	Annealed copper alloys, thin soft sheet metals
G	Ball, ∅ 1.588 mm	1.47	Malleable irons; phosphor bronze; gun-metal; cupro-nickel alloys, etc
H	Ball ∅ 3.175 mm	0.59	Soft materials; high ferritic aluminium, lead, zinc
K	Ball ∅ 3.175 mm	1.47	Aluminium and magnesium alloys
L	Ball ∅ 6.350 mm	0.59	Plastics } thermoplastic
M	Ball ∅ 6.350 mm	0.98	Plastics
P	Ball ∅ 6.350 mm	1.47	Plastics thermosetting
R	Ball ∅ 12.70 mm	0.59	Very soft plastics and rubbers
S	Ball ∅ 12.70 mm	0.98	—
V	Ball ∅ 12.70 mm	1.47	—

more gradual as grain growth occurs and the metal becomes fully annealed.

The effect of heating a quench-hardened plain-carbon steel is more gradual as shown in Fig. 11.31. During the tempering range of the steel no grain growth occurs, but there are structural changes. Initially, there is a change in the very hard martensite as particles of carbide precipitate

Table 11.6 Rockwell superficial hardness test conditions

Scale	Indenter	Additional force (kN)
15–N	Diamond cone	0.14
30–N	Diamond cone	0.29
45–N	Diamond cone	0.44
15–T	Ball, ∅ 1.588 mm	0.14
30–T	Ball, ∅ 1.588 mm	0.29
45–T	Ball, ∅ 1.588 mm	0.44

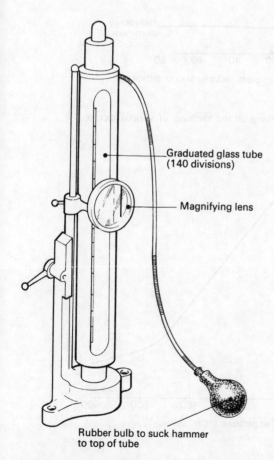

Graduated glass tube
(140 divisions)

Magnifying lens

Rubber bulb to suck hammer
to top of tube

Fig. 11.28 Shore scleroscope

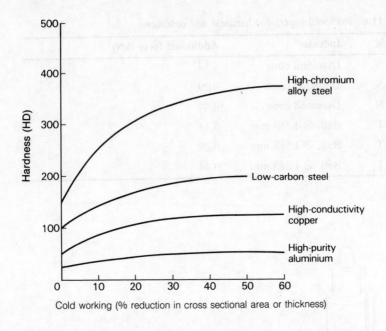

Fig. 11.29 Effect of cold-working on the hardness of various metals

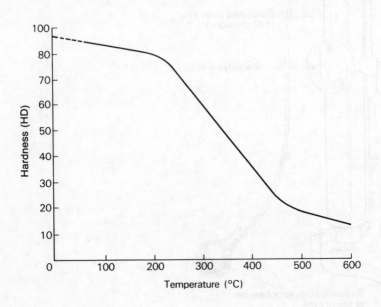

Fig. 11.30 Effect of heating cold-worked 70/30 brass

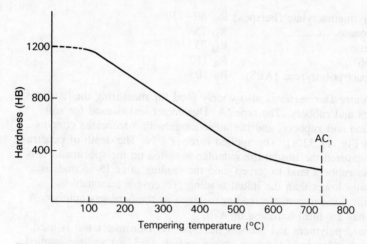

Fig. 11.31 Effect of heating a quench-hardened 0.8% plain carbon steel

out. As tempering proceeds and the temperature is increased, the structure loses its acicular martensitic appearance and spheroidal carbide particles in a matrix of ferrite can be seen under high magnification. (Spheroidising annealing was considered in Section 5.2.) These structural changes increase the toughness of the metal considerably, but with some loss of hardness.

11.13 The hardness testing of polymers

Brinell and Rockwell hardness tests may be used to test polymeric materials, but the size of the indenter and the applied load have to be varied to suit the softness of the material. For example, when using the Brinell test a 10 mm diameter hardened steel ball is used with a force of 4.9 kN applied for 30 seconds. Typical results are:

polystyrene	H_B 25	polyvinyl chloride	H_B 20
perspex	H_B 20	polyethylene	H_B 2

The Rockwell test is widely used for polymeric materials, particularly the M and the R scales (see Table 11.5). The M scale is used for the harder polymers and the R scale is used for the softer plastics and for the rubbers. Typical values are:

Melamine formaldehyde	R_M 120
Phenol formaldehyde	R_M 115
Urea formaldehyde	R_M 115
Polyester (not reinforced)	R_M 70–100
Epoxy	R_M 80–110

Polymethylmethacrylate (Perspex) R_M 80−110
Polycarbonate R_M 75
Polystyrene R_M 75
Nylon '66' R_R 110
High-impact polystyrene (ABS) R_R 105

The *Shore Durometer* is also widely used for measuring the hardness of plastics and rubbers. The type 'A' Durometer test is used for soft plastics and soft rubbers, and has an indenter with a truncated cone as shown in Fig. 11.32(*a*). The applied force is 8 N. The depth of penetration is measured the instant the indenter is seated on the specimen. This is because rubbers tend to 'creep' and the reading after 15 seconds is substantially lower than the initial reading. To ensure reasonable repeatability, the immediate reading is taken as the hardness number. A car tyre has a typical hardness of A70.

For hard polymers and rubbers the type 'D' Durometer test is used. This has an indenter of the type shown in Fig. 11.32(*b*) and the applied force is 44.5 N.

A quick workshop test is the *pencil test*. This is in fact a scratch resistance test. A set of pencils from 2B, B, HB, H, 2H, up to 9H are carefully sharpened to a wedge and dressed on abrasive paper. Each pencil, in turn, is held at right angles to the test piece surface and drawn across the material. The least hard pencil to leave a visible mark (indentation) on the surface of the plastic is reported as the hardness grade.

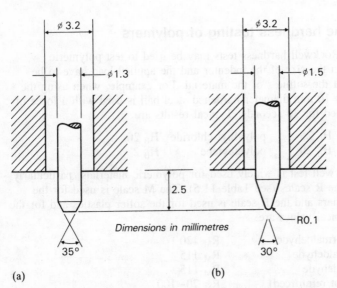

Fig. 11.32 Durometer indenters (*a*) Type 'A' indenter (*b*) Type 'D' indenter

As an alternative to hardness, polymeric materials may be tested for *softness* (BS 2782). This test involves a ball indenter 2.38 mm diameter being pressed into the surface of the plastic by an initial force of 5.25 N applied for 5 seconds, after which an additional force of 5.25 N is applied for a further 30 seconds. The difference between the penetration depths for the two forces is measured and expressed as the *softness number*. This number is the actual difference in depth expressed in units of 0.01 mm. Hence a test which gives an initial penetration of 0.1 mm and a final penetration of 0.5 mm, would have a difference in penetration of 0.4 mm, or a softness number of 4. The test is performed at a temperature of 23°C ± 1°C.

11.14 Comparative scales of hardness

Most hardness testing machine manufacturers issue tables showing comparative hardness numbers for various methods of testing. These tables should be treated with caution since the tests are carried out under different conditions. For instance the ball indenter of the Brinell test displaces the metal by plastic flow, whilst the diamond indenter of the Vickers test tends to cut its way into the test piece by shear, and the scleroscope is a dynamic test measuring hardness as a function of resilience.

Figure 11.33 shows the approximate relationship between typical hardness scales. These scales are related by a horizontal line across the figure. The Moh scale of hardness is included for comparison and is based upon a list of naturally occurring materials. These are arranged so that each material will scratch the material appearing immediately below it. Thus the hardest material (diamond) is placed at the head of the list, and the softest material (talc) is placed at the bottom of the list.

Table 11.7 lists some hardness values for a range of typical engineering materials.

11.15 Ductility testing

The percentage elongation, as determined by the tensile test, has already been discussed (Section 11.5) as a measure of ductility. Another way of assessing ductility is a simple bend test. There are several ways in which this test can be applied, as shown in Fig. 11.34. The test chosen will depend upon the ductility of the material and the severity of the test required.

Close bend test

The specimen is bent over on itself and flattened. No allowance is made for spring back, and the material is satisfactory if the test can be completed without the metal tearing or fracturing. This also applies to the following tests.

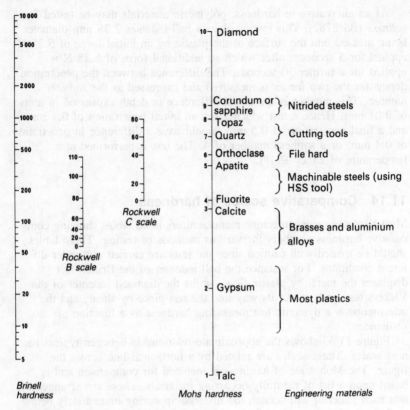

Fig. 11.33 Hardness scales

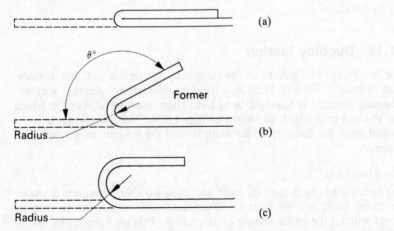

Fig. 11.34 Bend tests (a) Close bend test (b) Angle bend test (c) 180° bend test

Table 11.7 Hardness of typical engineering materials

Material	Condition	Hardness	
Aluminium bronze	Annealed	H_D	80
	Hard	H_D	220
Cartridge brass (70/30)	Annealed	H_D	65
	Hard	H_D	185
Commercial brass (63/37)	Annealed	H_D	65
	Hard	H_D	185
Muntz brass (60/40)	Extruded	H_D	75
Cupro–nickel (25% Ni)	Annealed	H_D	80
	Hard	H_D	170
Duralumin	Age hardened	H_D	30
Grey cast iron	—	H_D	210
Low tin bronze	Annealed	H_D	60
	Hard	H_D	210
Magnesium alloy	Wrought	H_D	65
Low carbon steel	Hot rolled	H_B	150
Medium carbon steel	Quench hardened and tempered at 550–650 °C	H_B	250
High carbon steel	Quench hardened and tempered at 150–300 °C	H_B	800
Stainless steel (ferritic)	Annealed	H_B	150
Stainless steel (austenitic)	Annealed	H_B	170
Stainless steel (martensitic)	Cutlery temper	H_B	534
	Spring temper	H_B	450
Polyvinyl chloride (rigid) (PVC)	—	H_{RR}	110
Polystyrene (high density)	—	H_{RM}	80
Polyamide (nylon '66')	—	H_{RR}	110
Acrylonitrite–butadiene –styrene (ABS)	—	H_{RM}	70

H_B = Brinell hardness H_R = Rockwell hardness.
H_D = Vickers diamond pyramid numeral

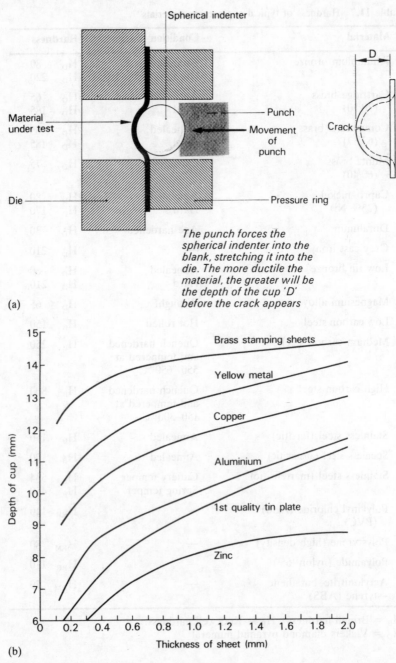

Spherical indenter

Material under test

Die

Punch

Movement of punch

Pressure ring

Crack

D

The punch forces the spherical indenter into the blank, stretching it into the die. The more ductile the material, the greater will be the depth of the cup 'D' before the crack appears

(a)

(b)

Fig. 11.35 Erichson cupping test (*a*) Principle of the cupping test (*b*) Erichson standard curves

Angle bend test

The material is bent over a former and the nose radius of the former and the angle of bend ($\theta°$) are fixed by specification. Again no allowance is made for spring back.

180° bend test

This is a development of the angle bend test using a flat former as shown. Only the nose radius of the former is specified.

Reverse bend test

This is used for very ductile materials where the preceding tests are not severe enough. The material is bent round a former of specified nose radius through 90° or 180° for a specified number of times.

In all the above tests the test piece is 10 mm wide and has rounded edges.

Another test widely used for testing the suitability of a sheet material for deep drawing in the press-shop (e.g. the cold pressing of car body panels) is the *Erichson cupping test*. This test simulates a deep drawing operation on a sample piece of material as shown in Fig. 11.35(*a*). The circular, sample blank is gripped between the pressure ring (blank holder)

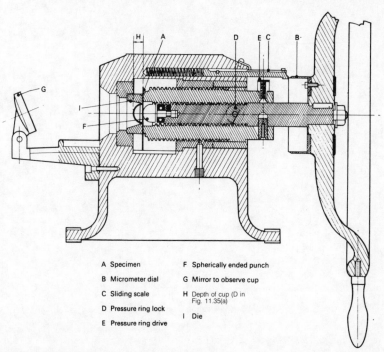

A Specimen	F Spherically ended punch
B Micrometer dial	G Mirror to observe cup
C Sliding scale	H Depth of cup (D in Fig. 11.35(a)
D Pressure ring lock	I Die
E Pressure ring drive	

Fig. 11.36 The Erichson cupping machine

312

and the die face to prevent the sample from puckering. After the pressure ring has been tightened down on the blank it is slackened back by a standard amount (0.05 mm) to give the metal freedom to draw. The spherically ended punch is then pressed into the sample blank, drawing it into the die. The forward movement of the punch continues until a crack is seen to appear at the base of the cupped sample. The depth of the drawn cup is read off the scale of the machine and is a measure of the drawing quality of the sample material. Figure 11.35(b) shows Erichson's standard curves giving typical values for depth of cup which may be expected for various thicknesses of material and for various materials. Figure 11.36 shows a section through an Erichson cupping machine. Bend tests and cupping tests are not absolute tests of ductility, but compare the performance of sample materials with materials of known performance.

12 Materials testing (non-destructive)

12.1 The need for non-destructive testing

The determination of material properties by destructive testing has already been discussed in Chapter 11. Although such tests are invaluable for the quality control of materials, such tests cannot control the quality of components, fabrications and assemblies made from these materials. For instance, a tensile test on a cast test piece may show the correct yield point, strength, and elongation percentage. It cannot show whether a casting made from the material contains blowholes, porosity and cold shuts. Further, the grain structure of the test piece will be unlike the grain structure of a large casting which will have cooled more slowly, therefore the properties of the test piece will not accurately reflect the properties of the casting. In the case of a hollow casting, destructive testing cannot show the effects of a misplaced or distorted core.

For safety, welds on pipe runs and pressure vessels must be free from discontinuities such as porosity and slag inclusions. Such welds must also have correct fusion and penetration. They should show no reduction in cross-sectional area due to lack of filler material. Again, for safety, the forged light-alloy hinges for the control surfaces of aircraft must be free from cracks due to forging at too low a temperature, or grain growth due to forging at too high a temperature.

A full programme of non-destructive testing can be very expensive, requiring each component to be handled individually through each series of tests. The tests themselves are often complex (for example radiography) requiring expensive equipment and skilled personnel in their execution.

For small components produced in batches and whose performance is not critical it is sufficient to take random samples from each batch, inspect them visually and, if hollow, section them to see that the cored holes are to specification. If the occasional blowhole turns up during machining, then the component can be rejected and melted down again. Sample forgings can be sectioned, etched, and given a macro-examination.

However, such sampling followed by destructive examination is not appropriate for large castings produced singly or in small batches. A fault lying undiscovered in a turbine housing or discovered only after much expensive machining has taken place would be disastrous not only in terms of human safety but financially as well. A faulty weld in the legs of an offshore oil-rig or an under-sea pipeline could have consequences out of all proportion to the cost of carrying out a full test of each weld. Similarly, each critical component of an aircraft must be fully and individually inspected. Obviously such tests and examinations must be non-destructive and in no way affect the properties and performance of the component or assembly.

The non-destructive testing techniques to be considered in this chapter are:

(a) visual examination;
(b) visual examination assisted by dye-penetrants;
(c) ultrasonic testing;
(d) eddy current testing;
(e) magnetic testing;
(f) radiography (X-rays and gamma-rays).

12.2 Visual examination

This is the simplest and cheapest possible non-destructive examination. A visual examination of a casting can identify such defects as surface cracks; scabs; surface porosity; warping and twisting; and inadequate filling of the mould.

Preliminary dimensional checks and marking-out will indicate whether or not sufficient machining allowance is present so that machined surfaces will 'clean-up'. Such checks can also determine whether cored holes are correctly positioned and of correct size, also whether such holes are correctly centred in their bosses.

In the case of welds, visual inspection can show whether or not there are any large surface cracks, inadequate fusion, lack of penetration, insufficient filler material, poor joint shape, surface porosity and undercuts. Figure 12.1 shows the appearance of two faulty welds which are easily identified by visual inspection.

12.3 Use of dye penetrants

The following techniques are used to make visible surface cracks which are so fine that they normally escape visual inspection. A very old technique, which is still widely used, is to immerse the casting in a bath of

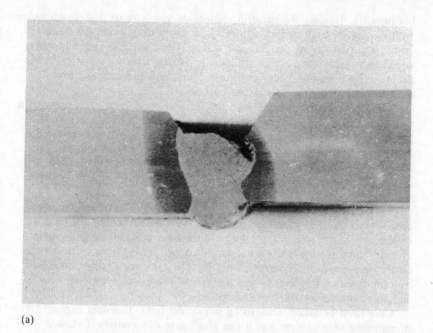

(a)

(b)

Fig. 12.1 Visual inspection of weld faults (*a*) A manual metal arc single 'V' butt weld, showing a lack of penetration and misalignment of the plate edges (*b*) A manual metal arc single 'V' butt weld, showing undercut and misalignment of the plate edges

hot paraffin. Heating the paraffin reduces its viscosity so that, combined with its already high surface-tension, it is easily drawn into the finest cracks and porosity by capillary attraction. The casting is removed and wiped thoroughly clean, after which it is painted with whitewash. Paraffin, exuding from the cracks, will discolour the whitewash and reveal the presence of surface cracks and porosity.

A more modern technique used for large castings is to paint them with special penetrants which are readily drawn into the surface cracks and porosity at room temperature. The penetrants contain brightly coloured dyes. After allowing time for the penetrant to be absorbed into any surface cracks which may be present, the casting is thoroughly cleaned off and dusted with a white powder. Coloured marks on the powder indicate where the cracks are.

A more sophisticated technique is to use specially developed fluorescent penetrants. The penetrant is painted over the surface of the casting or applied by pressure aerosol spray in the case of small castings. After allowing time for penetration the surplus penetrant is cleaned off and the casting is viewed under *ultraviolet* light. The presence of any penetrant exuding from surface cracks and porosity will appear as glowing lines and spots against a dark background.

Care must be taken when using ultraviolet light. Constant exposure to it can cause skin cancers and it can also cause eye injuries if viewed directly. These eye injuries are similar to the problem of 'arc-eye' which is caused when an arc welding flash is received when a welder is not wearing a suitable protective face visor with a filter-glass screen.

12.4 Ultrasonic testing

Although it varies widely from person to person, the highest frequency a normal person can be expected to hear is pitched at about 18 kHz. Sound waves of an even higher frequency can be heard by most animals, as their hearing is more acute than that of human beings. Sound waves whose frequency is too high to be heard by human beings are said to be *ultrasonic*. The higher the frequency (the shorter the wave length) of the ultrasonic sound waves, the easier it is to beam them and control them. The frequencies used for ultrasonic testing lie between 0.5 MHz and 15 MHz depending upon the material under test. Frequencies between 1 MHz and 3 MHz are suitable for steel components.

Pulses of high frequency oscillations are generated electronically, amplified, and fed into suitable transducers for conversion into sound waves. Such transducers exploit either (*a*) magnetostriction effects, or (*b*) piezoelectric effects.

Magnetostriction

This relies upon the fact that soft iron changes dimensionally in the presence of an alternating magnetic field. Thus a rod of soft iron placed in a coil, through which an alternating current is being passed, will increase and decrease in length in step with the alternations in the magnetic field and, therefore, in step with the alternating current producing that field. The oscillations in length drive a diaphragm which produces ultrasonic waves in the air.

Piezoelectric effects

These rely upon the fact that certain crystals such as quartz change dimensionally when small electric currents pass through them. Thus when an alternating current passes through the crystal, the crystal will change in size in step with the alternations. The 'bleep' signal in computers and electronic alarm clocks is usually generated using piezoelectric devices. An advantage of the piezoelectric effect is that it is reversible. Not only can an electric current change the size of the crystal, straining the crystal mechanically generates a corresponding electric current. This latter effect is exploited in record player pickups and in microphones. Thus a transducer using piezoelectric effects can be used both as an ultrasonic transmitter and as a receiver.

The basic principle of ultrasonic testing is shown in Fig. 12.2. The

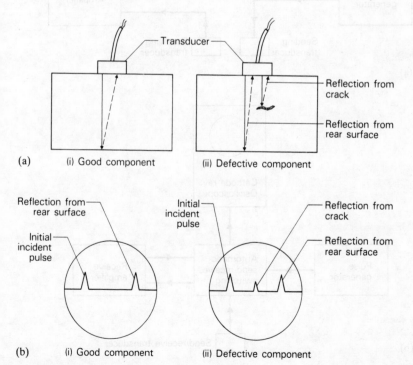

Fig. 12.2 Principles of ultrasonic testing (*a*) Ultrasonic testing (*b*) Typical oscilloscope displays

pulses of high frequency oscillations are driven into the component by the transducer. When the sound waves meet any discontinuity, such as a crack, the waves are reflected back into the transducer where they are converted into electrical pulses which can be displayed on the screen of an oscilloscope. By using pulses of short duration, the same transducer can transmit and receive alternately as shown in Fig. 12.2(a). The appearance of the display on the screen of an oscilloscope is shown in Fig. 12.2(b). Since the distance between the incident pulse and the

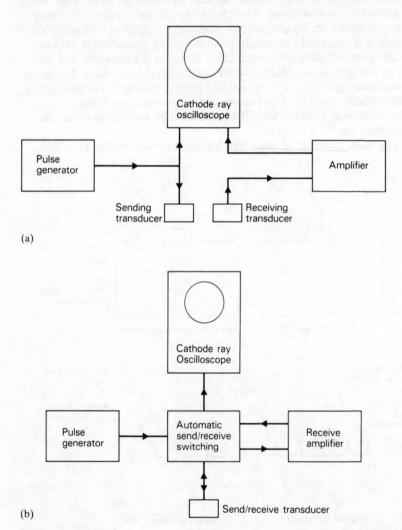

Fig. 12.3 Ultrasonic test equipment (a) Dual transducer equipment (b) Single transducer equipment

reflection from the back surface of the component is proportional to the component thickness, the position of the reflection from the fault clearly indicates where it lies in the component relative to the front and back faces. Figure 12.3 shows a block diagram of the complete equipment.

The use of a single, common transducer for transmission and reception will detect most randomly oriented faults and is quick and easy to use and interpret. However, it may miss a long, narrow fault whose axis is parallel to the path of the ultrasonic waves as shown in Fig. 12.4(a). To overcome this problem the more difficult technique of using separate transducers for transmission and reception may be employed as shown in Fig. 12.4(b). Obviously, it is much more difficult to position the fault by this technique.

The transducers must be in intimate contact with the surface of the component or false echoes will occur. This contact is achieved by placing a film of oil between the transducer(s) and the component surface, so that no air gap can exist. One great advantage of ultrasonic testing is that it can be equally well used on magnetic and non-magnetic metals, and on non-metals.

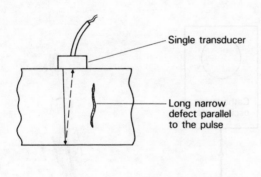

(a)

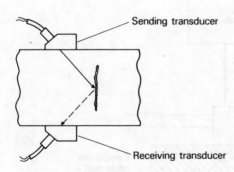

(b)

Fig. 12.4 Multiple transducer technique (a) A single transducer can miss a narrow defect parallel to the pulse (b) Defect detected by using dual transducers

320

12.5 Eddy current testing

This technique depends upon the fact that if an alternating magnetic field is linked with any metallic object, eddy currents will be induced in the metallic object by electro-magnetic induction. This applies to all metals whether they are ferromagnetic or non-magnetic. Figure 12.5 shows the principle of the test. A high-frequency alternating current is made to flow through a small coil and, as a result, a high frequency alternating magnetic field is set up around the coil. This field induces small electric currents called eddy currents which circulate in the component and produce magnetic fields in their own right. These secondary fields react with the field of the induction coil and affect the flow of current through that coil. Providing there is no change in section or composition as the induction coil is moved across the surface of the component the resultant current in the coil will remain constant. If, however, the induction coil moves over a crack or other fault in the component, the eddy current system will be temporarily disrupted and the resultant current in the coil will change. Therefore, if the induction coil is moved over the surface of the component and the resultant current through the coil is constantly

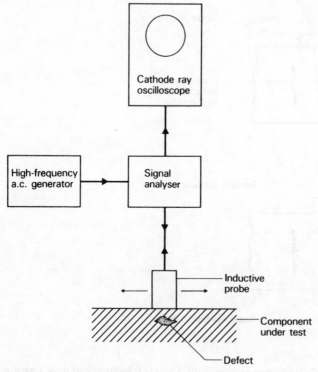

Fig. 12.5 Principle of eddy current testing

monitored on the screen of an oscilloscope, then any cracks or other discontinuities in the component will be revealed.

12.6 Magnetic testing

This method of crack detection, although simple and reliable, can only be applied to ferromagnetic materials. Further, it is only appropriate for surface cracks and discontinuities not more than 10 mm below the surface of the component. This technique is based upon the fact that the magnetic susceptibility in the region of a discontinuity is inferior to that of the surrounding metal and that this distorts the magnetic flux distribution. The resulting distortion of the flux field is usually detected by means of magnetic powder (magnetic iron oxide) in suspension in light machine oil or paraffin. The suspension is spread thinly over the surface of the component and the magnetic powder 'bunches' in the vicinity of the fault as shown in Fig. 12.6.

The component under test may be placed between the poles of a powerful magnet as shown in Fig. 12.27. This results in the magnetic field lying parallel to the surface of the component and is ideal for locating faults at right-angles to the flux field as shown in Fig. 12.8.

However, for long bars the faults usually lie parallel to the axis of the bar and would not be shown up by magnetisation in the plane of the axis as shown in Fig. 12.9(a). For long thin bars it is better to pass a heavy direct electric current through the bar from end to end as this sets up magnetic flux fields at right-angles to the current flow as shown in Fig. 12.9(b). Thus the flux field will again be perpendicular to the plane of the anticipated faults. Bars up to 4 metres long and 75 mm diameter can be tested in this manner. In this instance, the bars are immersed in the suspension and slowly rotated so that they can be examined all over.

Magnetic crack detection, like other non-destructive tests can be applied to finished components. It is effective in showing up surface cracks invisible to the unaided eye. Such faults are not uncommon in the

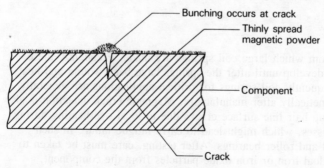

Fig. 12.6 Principle of magnetic crack detection

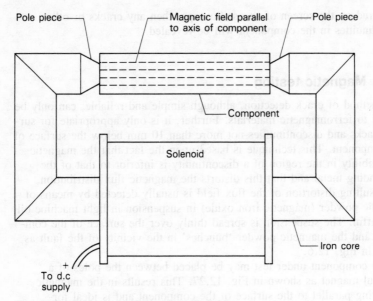

Fig. 12.7 Orientation of the magnetic field

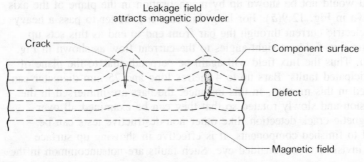

Fig. 12.8 Detecting faults perpendicular to the component surface

wire or rod from which large coil springs are made. Frequently these cracks do not develop until after the springs have been wound and heat-treated. Consequently coil springs for critical applications are generally inspected magnetically after manufacture. Magnetic crack detection is also used to show up hair line surface cracks due to local overheating during grinding processes, which might lead to early fatigue failure in such devices as ball and roller bearings. After testing, care must be taken to clean any residual iron or iron oxide particles from the component, particularly in the vicinity of bearing surfaces and screw threads.

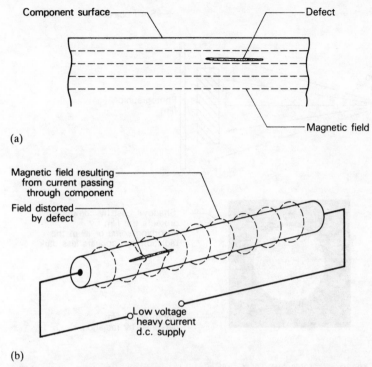

(a)

(b)

Fig. 12.9 Detecting faults parallel to component axis (a) Defects parallel to the magnetic field are not detected (b) Defects parallel to the component axis are perpendicular to the magnetic field when a direct current is passed through a component

12.7 Radiography

This is a photographic process in which the 'illumination' of the component is by X-rays or the even more penetrating gamma rays. These are electromagnetic radiations exactly the same as radio waves and light waves, except that they have a very much shorter wavelength (higher frequency). This enables X-rays and gamma rays to penetrate solid objects. When photographic film is exposed to X-rays or gamma rays and then developed, the film becomes dark. If a solid object is placed between the source of radiation and the film so that it casts a shadow on the film, the level of radiation reaching the film in the shadow area will be reduced and the shadow will appear on the film after development as a less dark area as shown in Fig. 12.10.

The thicker the material or the more dense it is, the greater will be the level of absorption. Therefore if there is a void in the material, the level of absorption will be lower at this point and the void will show up as a darker area on the film as shown in Fig. 12.11. The smallest defect

324

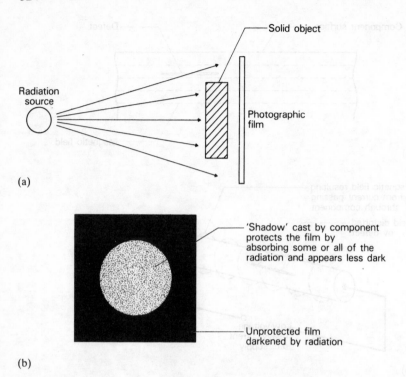

(a)

(b)

Fig. 12.10 Principle of radiography (*a*) General arrangement (*b*) Appearance of film after development

that can readily be detected by X-ray radiography is approximately 2 per cent of the section thickness. Below this the film cannot resolve the difference in radiation level. Thus for a component 25 mm thick any flaw under 0.5 mm in size will remain undetected. Thus the process becomes less reliable as the section thickness increases. In this respect it is at a disadvantage compared with ultrasonic testing which can detect small flaws in steel up to 15 metres thick with much less costly equipment and no potential radiation hazard. However, radiography has the advantage of providing a permanent visual record of the defect, its precise form, and its precise location.

X-rays are generated electrically in high-voltage discharge tubes. This method of generation provides a constant source of radiation, and the exposure time is easy to determine and control. Although widely used under laboratory or test department conditions, the size and weight of the equipment and the need for an electrical power supply makes it unsuitable for site work. It must be remembered that X-ray machines for examining metal components have to be much more powerful than those used for examining living tissues. Figure 12.12 compares the *half-value thicknesses* for mild steel using X-ray and gamma-ray sources. Half-value

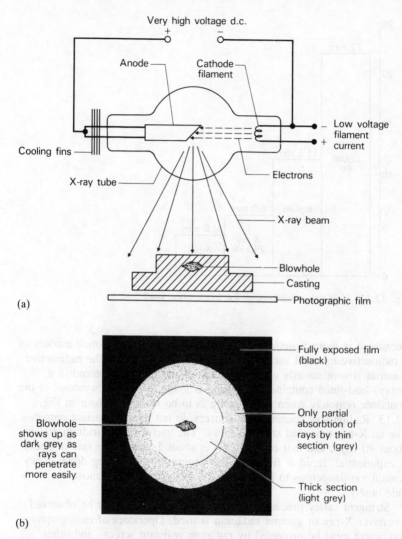

Fig. 12.11 Use of radiography to detect a casting defect (*a*) General arrangement (*b*) Appearance of film after development

thickness is the thickness of the material under examination which will reduce the intensity of the radiation to one half its incident value. The range of material thicknesses which can be examined by any given radiation source extends from one half-value thickness to about eight half-value thicknesses.

Gamma-ray sources are usually used for site work and thicker materials. This is because the radiation source is more portable and

326

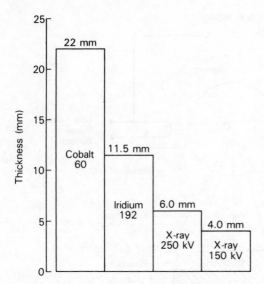

Fig. 12.12 Half-value thicknesses for low-carbon steel

because it is more penetrating. The radiation source is a small amount of a radioactive material such as iridium 192 or cobalt 60. The radioactive material is continuously emitting radiation and has to be stored in a heavy, lead-lined container. Provision is made to open a 'window' in the container remotely when an exposure is to be made as shown in Fig. 12.13. Radioactive (gamma-ray) sources do not give a constant output like an X-ray tube, but steadily decay. The useful life of iridium 192 is about 40 days, and that of cobalt 60 is about 5 years. The rate of decay is exponential, rapid at first and then progressively slowing down. They remain too dangerous to handle outside their containers almost indefinitely.

Stringent safety precautions and codes of practice must be observed whenever X-ray or gamma radiation is used. Operators of radiography equipment must be protected by radiation resistant screens and other persons must be kept well away from the radiation zone. Good ventilation is essential and the operator should change his or her protective overalls regularly. Normally the operator wears a radiation-sensitive tab which changes colour if exposed to excessive radiation but below the danger level to the operator. Further, radiographers should be subjected to regular medical checks. Gamma radiation sources must be stored in containers which are not only radiation resistant, but which are strong enough to resist mechanical damage if involved in an accident so that there is no leakage hazard.

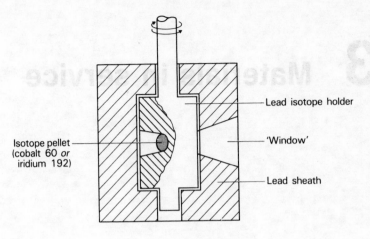

Lead isotope holder

'Window'

Isotope pellet
(cobalt 60 *or*
iridium 192)

Lead sheath

(a)

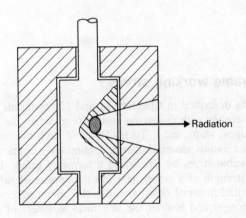

→ Radiation

(b)

Fig. 12.13 Typical γ-ray source (*a*) γ-rays 'Off' (*b*) γ-rays 'On'

13 Materials in service

13.1 Allowable working stress

Despite the tests described in Chapters 11 and 12, materials may still fail in service, sometimes with disastrous results (e.g. when the failure occurs in aircraft, bridges, ships, etc.). To try to avoid such disasters occurring the designer avoids materials being continuously operated at their maximum allowable stress by employing a *factor of safety*. Unfortunately, increasing the strength of a component in the interests of safety not only increases the initial material costs, but also the operating costs. For example the stronger and heavier the structural members of an aircraft, the fewer passengers it can carry and the more fuel it consumes. Therefore a balance has to be maintained between safety, initial cost and operating costs. The designer is constantly striving to improve the safety whilst reducing the cost.

Allowable working stress is taken as a proportion of the yield or proof stress, that is, the component is only stressed within its elastic range when in service. For example consider the screwed fastening shown in Fig. 13.1. When the nut is tightened normally the bolt is stretched slightly and, providing it is stressed within its elastic range, it will behave like a very powerful spring and it will pull the joint faces together very firmly. The stress in the bolt is made up of two elements. Firstly, that stress imparted by the initial tightening of the bolt, and secondly the stress imparted by the load on the fastening in service. The sum of these stresses must not exceed the yield stress for the material, and the designer proportions the fastening (and other components) so that there is

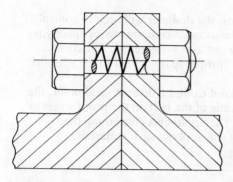

Fig. 13.1 The effect of stressing a bolt and nut.

a *factor of safety*. Usually the designer assumes an allowable working stress of only half the yield stress for the material.

If the bolt shown in Fig. 13.1 is over-stressed by applying excess torque to the nut, the bolt could well be stressed beyond the elastic range for the bolt material. This can occur if the spanner is extended with a piece of tube in an attempt to achieve greater tightness. In the case of a low-carbon steel bolt, for example, where the material has a distinct ductile yield point, the fitter will feel the bolt suddenly "give". The bolt will not fracture at this point but the sudden "give" indicates that the yield point has been exceeded and that the bolt has taken a permanent set.

Bolts, and other components, do deform plastically when stressed beyond the yield point for the material from which they are made. However, since the steel strain-hardens, the bolt still remains elastic to a higher stress than the initial elastic limit. This increased elastic strain is usually masked by the greater plastic strain.

Once the bolt is operating in the ductile region of the tensile stress-strain curve (see: Fig. 11.4) it will be subjected to both plastic and elastic strain. The plastic strain is *not* recoverable. Only elastic strain can be recovered when the bolt is unloaded.

In practice over-tightening the bolt results in the following sequence of events.

 (i) At the point of yield the bolt takes on a permanent set (becomes slightly and permanently lengthened) and the pre-louding of the joint interface is reduced.

 (ii) To try and re-establish this pre-loading the fitter takes advantage of the strain-hardening of the bolt and tightens it still further.

(iii) The additional work done on the bolt will now be shared between recoverable elastic strain which will be pre-loading the joint interface and non-recoverable plastic strain.

Not only will the clamping efficiency of the bolt be reduced, resulting in inadequate joint pre-loading, but it will be working dangerously near its fracture stress.

In an extreme case the bolt may fail in tension before the required magnitude of pre-loading is re-established.

Thus for critical assemblies, the designer will seek to control the stress in the fastening and associated components by specifying the **torque** to be applied to the nut as it is tightened. The correct torque is achieved by use of a "torque spanner" set to the specified value.

Where bolt material does not exhibit a distinct yield point, the designer bases the correct loading of the bolt on the proof stress or secant stress for the material. The arguments concerning the inadvisability of over-stressing still apply to such materials.

13.2 Creep

Creep is defined as the gradual extension of a material under a constant applied load. It is a phenomenon which must be considered in the case of metals when they are required to work continuously at high temperatures, for example, the blades of jet engines and gas turbines. Figure 13.2 shows a typical creep curve for a metal at high temperature. A constant tensile load is applied to a test piece in a tensile testing machine whilst the test piece is maintained at a constant elevated temperature. The creep curve obtained from this test shows three distinct periods of creep.

(a) *Primary creep.* This commences at a fairly rapid rate but slows down as work-hardening (strain-hardening) sets in and the strain-rate decreases. It can be seen from Fig. 13.2 that the extension due to creep is additional to the instantaneous elongation of material to be expected when any tensile load is applied (see Section 11.3). For calculating creep as a percentage elongation, the initial elongation is ignored and creep is considered to commence at point A on the curve.

(b) *Secondary creep.* During this period of creep the increase in strain is proportional to time. That is, the strain-rate is constant and at its lowest value.

(c) *Tertiary creep.* During this period of creep the strain rate increases rapidly, necking occurs and the test piece fails. Thus the initial stress, which was within the elastic range and did not produce early failure, did eventually result in failure after some period of time.

Creep in polymeric materials

Polymeric materials, however, are subject to creep effects even under normal ambient conditions. When used for engineering components, rigid polymers must be capable of withstanding reasonable tensile and compressive forces over long periods of time without dimensional change. Figure 13.3 shows some typical creep curves for a cellulose acetate material at 25°C. As is to be expected, the creep rate is highest for large applied loads and almost negligible for small applied loads, after the initial deformation has taken place.

Creep, for all materials, is calculated in the same manner as elongation, see tensile testing, that is:

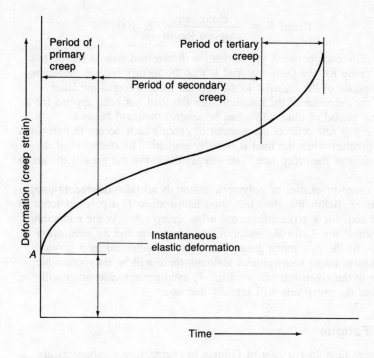

Fig. 13.2 Creep

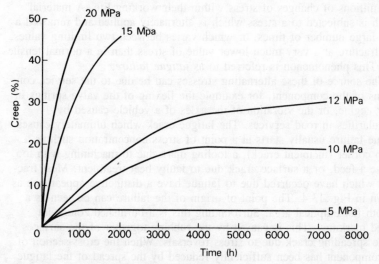

Fig. 13.3 Typical creep values (cellulose acetate)

$$\text{Creep \%} = \frac{\text{elongation}}{\text{original length}} \times 100$$

The difference between the elongation determined from a tensile test and the creep for the same material is that the former reflects the immediate response of the material to the applied load, whereas the latter reflects the response of the material after the load has been applied for a very long period of time. (This can be several thousand hours.)

The *creep rate* reflects the amount of creep which occurs in unit time. This is greatest when the load is initially applied. The steepness of the curve indicates the creep rate. The steeper the curve the greater the creep rate.

The creep properties of polymeric materials are also dependent upon temperature. Below the glass transition temperature T_g a polymer tends to be rigid and, for a given stress, has a low creep rate. As the temperature rises towards the T_g for the material, the creep rate and the elongation increase. At the T_g a much greater elongation is achieved by a given stress than at lower temperatures although there will be no appreciable increase in the creep rate. Beyond the T_g even greater elongation will occur, but the creep rate will actually decrease.

13.3 Fatigue

Since more than 75 per cent of failures in engineering components are attributed to fatigue failure, and as the performance from engineering products is continually increased, the need to understand the failure of materials from fatigue becomes increasingly important.

In service many engineering components undergo between thousands and millions of changes of stress within their working life. A material which is subjected to a stress which is alternately applied and removed a very large number of times, or which varies between two limiting values, will fracture at a very much lower value of stress than in a normal tensile test. This phenomenon is referred to as *fatigue failure*.

The source of these alternating stresses can be due to the service conditions of the component, for example the flexing of the valve springs in a car engine, or the vibration of the axles of a vehicle caused by irregularities in road services. The fatigue crack which ultimately causes fatigue failure usually starts at a point of stress concentration such as a sharp corner (incipient crack), a tooling mark due to machining with too coarse a feed, or a surface crack due to faulty heat treatment. Most fractures which have occurred due to fatigue have a distinctive appearance as shown in Fig. 13.4. The point of origin of the failure can be seen as a smooth flat elliptical area. Surrounding this is a burnished zone with ribbed markings. This is caused by the rubbing together of the surfaces of the spreading crack due to stress reversals. When the cross-section of the component has been sufficiently reduced by the spread of the fatigue crack, the component will no longer be able to carry its designed load and will fail suddenly, leaving a crystalline area visible as shown.

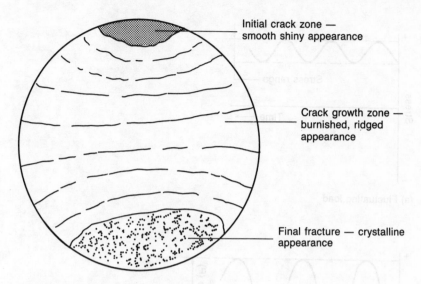

Fig. 13.4 Appearance of fatigue fracture

Three conditions of loading may occur and these will now be described.

(a) *Fluctuating load.* This varies between two positive limits as shown in Fig. 13.5(a).

(b) *Pulsating load.* This is also known as a 'repeated load' and applies to a load which fluctuates between zero and a positive value as shown in Fig. 13.5(b).

(c) *Alternating load.* This is also known as 'reversed loading'. The load fluctuates between a positive maximum load and a negative maximum load repeatedly as shown in Fig. 13.5(c).

The load may be applied in various ways. One of the most popular tests is to apply an alternating load in the bending mode as shown in Fig. 13.6(a). The test piece is supported in a rotating chuck at one end and loaded as a cantilever by a dead weight at the other. As the specimen rotates, the stress at any point on the circumference of the specimen alternates between tension and compression once during each revolution. To reduce the variation in applied stress from the centre (neutral axis of bending) to the outside of the specimen, it is sometimes made hollow. Alternatively, 'four-point loading' can be used as shown in Fig. 13.6(b). This loading gives a pure bending moment on the centre portion of the specimen and eliminates the shearing stress that exists when the cantilever specimen is used. The testing machine is fitted with a revolution counter to count the number of stress reversals, and a trip to stop the machine when failure of the specimen occurs.

If a large number of tests are made on a correspondingly large

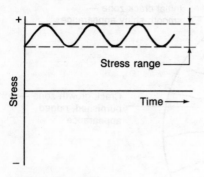

(a) Fluctuating load

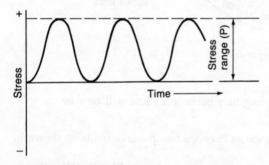

(b) Pulsating or repeated load

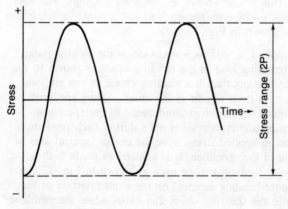

(c) Alternating load

Fig. 13.5 Conditions of fatigue loading

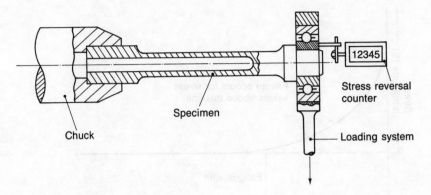

(a) Cantilever loading — alternating stress

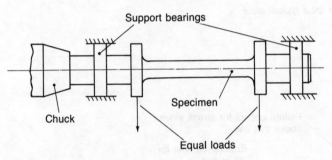

(b) Beam loading — alternating stress

Fig. 13.6 Fatigue testing

number of specimens of the same material, and if the stress range (2P, see Fig. 13.5) is gradually reduced, the number of reversals necessary to produce failure can be plotted against the range of stress. The graph produced will be similar to that shown in Fig. 13.7(a). This is called an *S-N diagram*. It implies that failure will never occur provided the range of stress is kept within the value of 'S_D', which is called the *fatigue limit*. From the nature of the test it is not possible to prove this and, in practice, the limiting range of stress which will not produce failure in 10×10^6 reversals is generally taken as the fatigue strength of the material. For some materials 'fatigue limit' values are not applicable as shown in Fig. 13.7(b). For such materials an endurance limit S_N is quoted. This defines the maximum stress amplitude which can be sustained for N reversals. Non-ferrous alloys usually have an S-N diagram similar to Fig. 13.7(b) and, since non-ferrous alloys are used in the manufacture of aircraft, this accounts for the reason why critical components have to be replaced after a specified number of flying hours (and

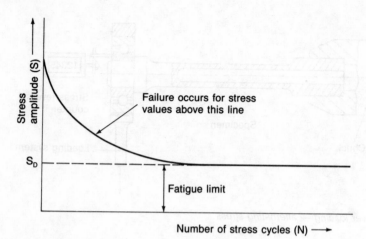

(a) S-N diagram for a typical steel

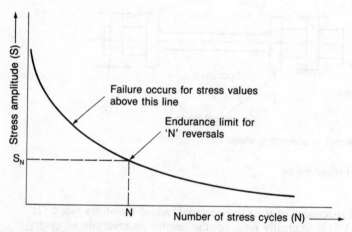

(b) S-N diagram for a typical non-ferrous alloy

Fig. 13.7 Stress reversal curves (S-N diagrams)

therefore stress reversals) even when they appear — visually — to be in good condition. The results depend upon the type of machine used and the type of test used, but similar test conditions give similar results. A typical S-N curve for a wrought aluminium alloy of the 'duralumin' type is shown in Fig. 13.8.

13.4 Factors affecting fatigue (metals)

Many factors affect the fatigue resistance of metals but the most important are given below.

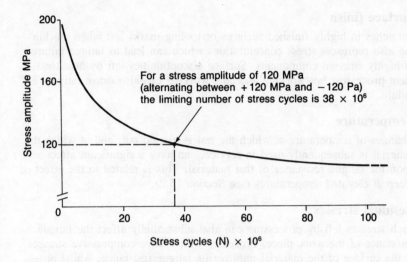

Fig. 13.8 S-N diagram for a wrought aluminium alloy

Design

Stress concentrations caused by sharp corners, sudden changes of section, or undercuts are all classified as 'incipient cracks' from which a fatigue crack may spread. To prevent such cracks spreading, round 'port holes' were used in ships' hulls long before fatigue failure was fully understood. Figure 13.9 shows how a small hole drilled through a steel component to cause a stress concentration can reduce the S_N value of the component by approximately a third.

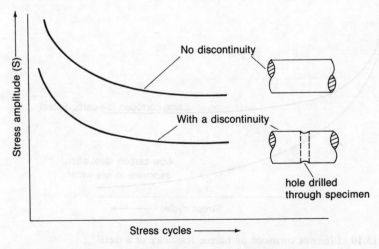

Fig. 13.9 Effect of a discontinuity on fatigue behaviour for a metal

338

Surface finish

Scratches in highly finished surfaces or tooling marks left when machining also represent stress concentrations which can lead to fatigue failure in highly stressed components. Surface discontinuities left by heat treatment processes, hot-working and cold-working can also cause fatigue failure.

Temperature

Changes of temperature at which the test is carried out, and at which the material is subsequently used in service, can have a significant affect upon the fatigue resistance of that material. This is related to the effect of creep at elevated temperatures (see Section 13.2).

Residual stresses

Such stresses left by processing can also substantially affect the fatigue resistance of the work piece. Processes which leave compressive stresses in the surface of the material improve its fatigue resistance, whilst processes which leave tensile stresses in the surface of the material reduce its fatigue resistance. In the latter case a stress relief treatment such as normalising can remove the internal stresses.

Corrosion

This may be atmospheric corrosion, oxidation during heat treatment or saline attack due to marine environments. Figure 13.10 compares the S-N curves for a plain carbon steel before and after exposure to sea-water. It can be seen that prior to exposure to sea-water the curve showed a clear fatigue limit (S_D) below which fatigue would not occur, whilst after exposure the steel showed no clearly defined fatigue limit and an

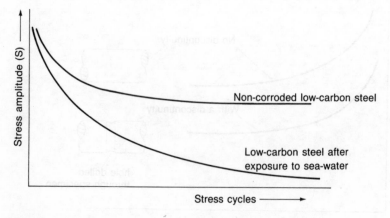

Fig. 13.10 Effect of corrosion on fatigue resistance of a metal

'endurance limit' would need to be specified. Surface treatment by galvanising and painting prior to saline exposure would prevent corrosion and would result in normal fatigue characteristics for the material.

13.5 Factors affecting fatigue (polymers)

Temperature

This has a much greater effect on polymeric materials than it has on metals (see Section 13.2) and substantially reduces the fatigue resistance of the material. Therefore temperature must be taken into account not only during the fatigue test but also during the service conditions in which the material must operate. Further, polymeric materials heat up internally when subjected to rapidly alternating stresses due to poor heat conducting properties within the polymer. The greater the stress and the more rapidly it alternates, the greater will be the temperature rise. Since this temperature rise results in a lowering of the fatigue resistance of the material, a limitation of the frequency of alternation must be placed on the test providing it is not reduced below the frequency of any stress alternation the material would be subjected to in service. Figure 13.11 shows a typical S-N diagram for unplasticised PVC.

Environment

The fatigue resistance of polymeric materials along with their general strength and toughness can be adversely affected by the presence of organic substances such as detergents, soaps and alcohols. 'Solvent cracking' also affects the fatigue resistance of polymers when they are exposed

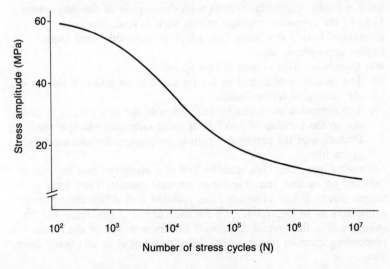

Fig. 13.11 S-N diagram for unplasticised PVC

to the fumes of such substances as toluene, acetone, benzene, etc. It is thought that the solvent is absorbed into the polymer chains under stress causing them to straighten and slip more easily.

Strong sunlight (UV radiation) and the presence of ozone can also cause some polymers to degrade and fail in fatigue at a lower than normal value (see Section 13.14).

13.6 The corrosion of metals

Corrosion is the slow but continuous eating away of metallic components by chemical or electro-chemical attack. That this is costly and destructive will be vouched for by any motorist who has seen his valuable 'pride and joy' eaten away before his eyes as body rot sets in. Three factors govern corrosion.

(a) the metal from which the component is made;
(b) the protective treatment the component surface receives; and
(c) the environment in which the component is kept.

All metals corrode to a greater or lesser degree; even precious metals like gold and silver tarnish in time and this is a form of corrosion. Corrosion prevention processes are unable to prevent the inevitable failure of the component by corrosion, but they do slow down the process to a point where the component will have worn out or been discarded for other reasons before failing due to corrosion. There are three ways in which metals corrode.

(a) *Dry Corrosion*. This is the direct oxidation of metals which occurs when a freshly cut surface reacts with the oxygen of the atmosphere. Most of the corrosion resistant metals such as lead, zinc and aluminium form a dry oxide film which protects the metal from further atmospheric attack.

(b) *Wet Corrosion*. This occurs in two ways.
   (i) The oxidation of metals in the presence of air and moisture as in the rusting of ferrous metals.
   (ii) The corrosion of metals by reaction with the dilute acids in rain due to the burning of fossil fuels (acid rain), for example the formation of the carbonate 'patina' on copper, the characteristic green film.

(c) *Galvanic Corrosion*. This occurs when two dissimilar metals, such as iron and tin or iron and zinc are in intimate contact. They form a simple electrical cell in which rain, polluted with dilute atmospheric acids, acts as an electrolyte. An electric current is generated and circulates within the system. Corrosion occurs with one of the metals (depending upon its position in the electrochemical series) being eaten away.

Table 13.1 Rate of corrosion

Type of environment	Typical rate of rusting for low-carbon steel in temperate climates (mm/year)
Rural	0.025–0.050
Urban	0.050–0.100
Industrial	0.100–0.200
Chemical	0.200–0.375
Marine	0.025–0.150

Fortunately the reactivity of a metal and the rate at which it corrodes are not related. For example, although aluminium is chemically more reactive than iron, as soon as it is exposed to the atmosphere it forms an oxide film which seals the surface and prevents further corrosion from taking place. On the other hand, iron is less reactive and forms its oxide film more slowly. Unfortunately the iron oxide film (rust) is porous and the process continues unabated until the metal is destroyed. Table 13.1 indicates the average rates of corrosion for unprotected steelwork in various environments.

13.7 Atmospheric corrosion

The cost of corrosion and its prevention is largely related to atmospheric corrosion, and it is this form of corrosion which will be considered in this chapter. The corrosion prevention problems associated with chemical engineering, marine engineering, food processing, etc., is more specialised and the remedies resorted to after a careful study of the environmental and service conditions are, correspondingly, more specific.

Any metal exposed to normal atmospheric conditions becomes covered with an invisible, thin film of moisture. This moisture film is invariably contaminated with dissolved solids and gases which increase the rate of corrosion. The most common example of corrosion due to dissolved oxygen from the atmosphere is the rapid surface formation of 'red rust' on unprotected ferrous metals. This 'red rust' is an oxide of iron, but a different oxide to the blue-black 'mill-scale' formed by heating in dry air. Figure 13.12 shows that air (oxygen) and moisture must both be present for rusting to occur. If either the air or the moisture is removed rusting cannot commence.

Once 'rusting' commences the action is self-generating, that is, it will continue even after the initial supply of moisture and air is removed. This is why all traces of rust must be removed or neutralised before painting, otherwise rusting will continue under the paint causing it to blister and

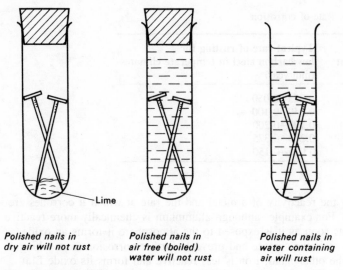

Polished nails in
dry air will not rust

Polished nails in
air free (boiled)
water will not rust

Polished nails in
water containing
air will rust

Fig. 13.12   Corrosion of steel

flake off. The rate of rusting slows down as the thickness of the rust layer increases, but as rain washes off the surface film of rust the process speeds up again. The cycle is continuous and, once started, is difficult to control.

Atmospheric pollution rapidly increases the rate of rusting of iron and steel. It also attacks, but more slowly, corrosion resistant metals such as copper and zinc. Lead is virtually unaffected and so is pure aluminium providing it has been correctly pre-treated and is regularly washed clean. How the gases given off by burning fossil fuels and by industrial processes are converted into corrosive substances is shown in Fig. 13.13. Gases which become acids when dissolved in water are called *anhydrides*. In coastal areas the problem is aggravated by the presence of salt spray in the atmosphere.

The most important gaseous pollutant is sulphur dioxide. Many 'clean' or 'smokeless' fuels are guilty of producing large quantities of this gas. Unfortunately the 'clean air' legislation only controls the visible products of combustion and ignores the fact that many of the most damaging pollutants are colourless gases which are invisible to the naked eye. It has been estimated that approximately 10 million tonnes of sulphurous and sulpuric acids are produced in the United Kingdom annually by the combustion of coal and coal products alone. This is about four times the amount of such acids produced intentionally.

13.8   Galvanic corrosion

It has already been stated that when two dissimilar metals come into intimate association in the presence of an electrolyte that a simple elec-

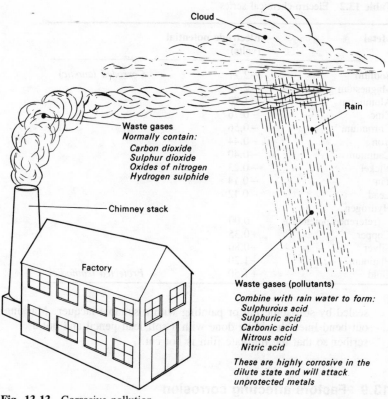

Cloud

Rain

Waste gases
Normally contain:
Carbon dioxide
Sulphur dioxide
Oxides of nitrogen
Hydrogen sulphide

Chimney stack

Factory

Waste gases (pollutants)

Combine with rain water to form:
Sulphurous acid
Sulphuric acid
Carbonic acid
Nitrous acid
Nitric acid

These are highly corrosive in the
dilute state and will attack
unprotected metals

Fig. 13.13 Corrosive pollution

trical cell is formed resulting in the eating away of one or other of the metals. Metals can be arranged in a special order called the *electro-chemical series*. This series is listed in Table 13.2 and it should be noted that, in this context, hydrogen gas behaves like a metal. If any two metals come into contact in the presence of a dilute acid, the more negative metal will corrode more rapidly and will be eaten away. Figure 13.14 shows what happens during electrolytic corrosion. For example:

(a) In the case of galvanised iron (zinc-coated mild steel) the zinc corrodes away whilst protecting the steel. The zinc is said to be *sacrificial*. The destruction of the zinc coating can be retarded by the application of a protective paint film. The zinc forms an excellent 'key' for the paint, and the paint film keeps the wet or moist corrosive atmosphere away from the zinc and prevents any galvanic action occurring.

(b) In the case of tin-plate, the mild steel base is corroded away if the coating is broken at any point. Hence cut edges should always be

Table 13.2 Electro-chemical series

Metal	Electrode potential (volts)	
Sodium	−2.71	*Corroded (anodic)*
Magnesium	−2.40	
Aluminium	−1.70	
Zinc	−0.76	
Chromium	−0.56	
Iron	−0.44	
Cadmium	−0.40	
Nickel	−0.23	
Tin	−0.14	
Lead	−0.12	
Hydrogen (reference potential)	0.00	
Copper	+0.35	
Silver	+0.80	
Platinum	+1.20	
Gold	+1.50	*Protected (cathodic)*

sealed by soft-soldering or painting with a suitable lacquer. Marking out bend lines should be done with a soft lead pencil and not a scriber so that the tin-plate film is not cut.

13.9 Factors affecting corrosion

Structural design

The following factors should be observed during the design stage of a component or assembly to reduce corrosion to a minimum.

(a) The design should avoid crevices and corners where moisture may become trapped, and adequate ventilation and drainage should be provided.
(b) The design should allow for easy washing down and cleaning.
(c) Joints which are not continuously welded should be sealed, for example by the use of mastic compounds or impregnated tapes.
(d) Where dissimilar metals have to be joined, high strength epoxy adhesives should be considered since they insulate the metals from each other and prevent galvanic corrosion.
(e) Materials which are inherently corrosion resistant should be chosen or, if this is not possible, an anti-corrosive treatment should be specified.

Environment

The environment in which the component or assembly is to spend its service life must be carefully studied so that the materials chosen, or the

345

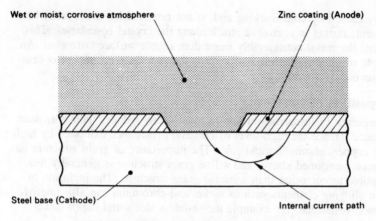

Wet or moist, corrosive atmosphere   Zinc coating (Anode)

Steel base (Cathode)   Internal current path

(a) Protection by a sacrificial coating

Coating is eaten away whilst protecting the base

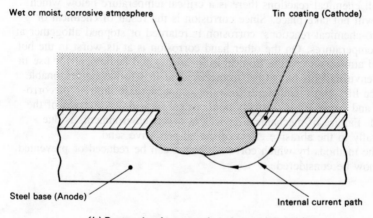

Wet or moist, corrosive atmosphere   Tin coating (Cathode)

Steel base (Anode)   Internal current path

(b) Protection by a purely mechanical coating

Coating only protects the base if intact.
If coating is damaged, base is eaten away quicker than if coating were not present.

Fig. 13.14 Electrolytic corrosion

anti-corrosion treatment specified, will provide an adequate service life at a reasonable cost. It would be unnecessary and uneconomical to provide a piece of office equipment which will be used indoors with a protective finish suitable for heavy-duty contractors' plant which is going to work on construction sites in all kinds of weather conditions.

Applied or internal stresses

Chemical and electro-chemical corrosion is intensified when a metal is under stress. This applies equally to externally applied and internal stresses, although more common in the latter case. Internal stresses are usually caused by cold-working and, if not removed by stress-relief heat treatment, results in corrosive attack along the crystal boundaries. This weakens the metal considerably more than simple surface corrosion. An example of inter-crystalline corrosion is the 'season cracking' of α brass after severe cold working.

Composition of structure

The presence of impurities in non-ferrous metals reduces their corrosion resistance. Hence the high level of corrosion resistance exhibited by high purity copper, aluminium and zinc. The importance of grain structure has also been mentioned above, and a fine grain structure is generally less susceptible to corrosion than a coarse grain structure. The inclusion of certain alloying elements such as nickel and chromium can also improve corrosion resistance, for example the stainless steels and cupro-nickel alloys.

Temperature

For all chemical reactions there is a critical temperature below which they will not take place. Since corrosion is the result of chemical or electro-chemical reactions, corrosion is retarded or stopped altogether at low temperatures. On the other hand corrosion is at its worst in the hot, humid atmosphere of the tropical rain forests, and equipment for use in such environments has to be 'tropicalised' if it is to have a reasonable service life. High temperatures alone do not increase the rate of corrosion, and corrosion is virtually non-existent in arid desert areas of the world. Failure of mechanical devices in desert environments is due generally to the abrasive effect of the all-pervasive sand.

The methods by which corrosive attack can be reduced or prevented will now be considered.

13.10 Metals which resist corrosion

It has already been made clear that metals combine with atmospheric oxygen and/or atmospheric pollutants to a greater or lesser extent. The following metals, which resist corrosion, react to form impervious, homogeneous coatings on their surfaces which prevent further corrosion from taking place, providing these coatings remain undisturbed.

Copper

When copper is exposed to atmospheric pollution for a long time, as for example when used as a roofing material, the surface develops a green coating or 'patina'. The 'patina' is caused by the action of the acids in the atmosphere attacking the basic oxide coating of the copper to form a protective layer of sulphate and carbonate salts as shown in Fig. 13.15. Atmospheric acids do not attack the copper directly.

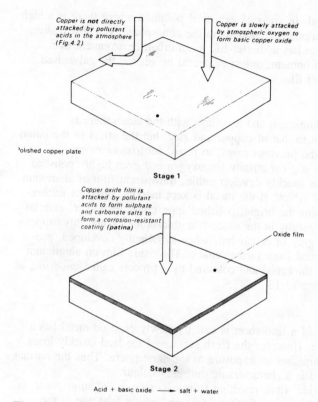

Copper is **not** directly
attacked by pollutant
acids in the atmosphere
(Fig 4.2)

Copper is slowly attacked
by atmospheric oxygen to
form basic copper oxide

Polished copper plate

Stage 1

Copper oxide film is
attacked by pollutant
acids to form sulphate
and carbonate salts to
form a corrosion-resistant
coating (patina)

Oxide film

Stage 2

Acid + basic oxide ⟶ salt + water

Fig. 13.15 Formation of 'Patina' on copper

The 'patina' which forms on copper, giving it a protective skin, must not be confused with the darker green compound 'verdigris' which forms on copper by reaction with organic matter. Verdigris corrosion will, in time, completely destroy copper. Thus copper vessels used in the food processing industry and for cooking, are heavily coated with tin to prevent verdigris forming.

Zinc

The results of atmospheric attack upon zinc is very similar to copper, especially for outdoor purposes. In this instance it forms a protective carbonate coating which strengthens with time. This coating is grey in colour, not unlike the parent metal itself, and does not crack or peel off with any expansion or contraction of the metal due to temperature changes. For this reason zinc is an excellent exterior building material and is widely used for flashings and roof cladding. Similarly it gives excellent protection when coated onto steel (see galvanising, Section 13.12). Its useful life tends to be foreshortened, particularly as a galvanising material, in areas of industrial pollution where there is a high concentration of sulphur compounds in the atmosphere. For this reason, galvanised hardware has a shorter life in an urban environment than it has in a rural environment, unless protected by sealing the galvanised surface with a paint film.

Aluminium

The reaction of aluminium and its alloys with the atmosphere is somewhat different to that of copper and zinc, but the effect of the initial reaction is, as in the previous cases, to retard further corrosion.

Aluminium has a great affinity for oxygen and even highly polished aluminium surfaces quickly develop a thin, transparent film of aluminium oxide or 'alumina' which, if the metal is kept indoors, prevents further oxidation and retains the bright, polished appearance. However, exterior use of aluminium results in the oxide film thickening. When this happens the film becomes grey in colour but, when sufficiently developed, protects the parent metal from further attack. The oxide film on aluminium can be artificially thickened and coloured by a process called *anodising* as described in Section 13.13.

Lead

When the surface of a lead sheet is cut, the newly exposed metal has a silvery appearance. However the fresh surface of the lead quickly loses its lustre which tarnishes on exposure to the atmosphere. Thus the normal appearance of lead is a characteristic dull grey colour.

This 'white oxide' film, resulting from exposure to the atmosphere, is very tenacious and prevents further attack and makes lead one of the most corrosion resistant of all metals. It has the property of remaining incorrodible for year after year and is widely used as an extruded, protective sheathing for underground telephone and power cables. It is also used as a protective metal coating for sheet iron and steel — 'terne-plate'. It is no longer used for water pipes because it is slightly soluble in water and its toxic effects on the body are cumulative.

Stainless steel

Stainless steels are outstanding corrosion resistant metals. Unlike the metals considered so far in this section, stainless steels have high structural strength and high corrosion resistance even at elevated temperatures which would melt other corrosion resistant metals. One familiar stainless steel is known as '18/8' because the alloy contains 18 per cent chromium and 8 per cent nickel, the remainder being iron.

As with aluminium and lead, it is the formation of a complex oxide film which protects the surface from attack. The grades containing the additional alloying element molybdenum are recommended for architectural applications in heavily polluted areas as they will retain their polish and remain unchanged in appearance indefinitely. The presence of molybdenum also reduces the likelihood of 'weld-decay' in fabricated stainless steels.

Stainless steels are also widely used in chemical plant and food processing and handling equipment where they combine corrosion resistance to a wide range of substances at elevated temperatures, high strength, and non-toxic properties. Some grades can also be hardened and sharpened to a cutting edge for cutlery. No finishing treatment is required by stainless steels other than polishing or brushing. Although expensive in initial cost, it is economical in the long term as maintenance is minimal.

Nickel

This metal has already been mentioned as a constituent of stainless steels. It is also used for 'electroplating' onto other metals to provide a protective coating. It also forms corrosion resistant alloys with copper and these were discussed in Section 7.11.

Chromium

This metal has already been mentioned as a constituent of stainless steel. It is also used for 'electroplating' but, unlike nickel which can be deposited directly onto a variety of base metals, chromium is used as a finishing film over a nickel plated sub-strata. It has a more pleasing colour than nickel and does not tarnish.

13.11 Chemical inhibition

When unpacking a new set of slip gauges or similar highly finished equipment, a sheet of VPI paper is often found in contact with the metal. These letters stand for 'Vapour Phase Inhibitor' and the paper is impregnated with chemical substances which give off a vapour neutralising any corrosive atmosphere which may remain within the packing or which may penetrate the packing. Any residual moisture, which could form an electrolyte and cause electrochemical corrosion, being rendered alkaline so that the metal becomes passive.

13.12 Cathodic protection

The protection of iron by galvanising has already been introduced. The zinc is more corrosion resistant than the iron and is less likely to be corroded. In the event of corrosion occurring, the zinc is eaten away rather than the iron or steel. This is because the iron is electro-positive compared with zinc. That is, the iron forms the *cathode* of the electrolytic cell which exists when iron and zinc are in contact in a corrosive environment. Thus the use of zinc to protect ferrous metals is called *cathodic protection*. Microporous zinc electrodes are often fitted to the steel hulls of ships adjacent to the bronze propellors. This prevents electrolytic attack of the hull (the highly saline sea water being the electrolyte) and the easily replaced zinc block is eaten away instead. This is cathodic protection and the zinc is said to be *sacrificial*.

13.13 Protective coatings

Where the natural protective film which builds up on the surface of a metal is inadequate, or where (as in the case of most ferrous metals) that film is inadequate, additional protective coatings have to be applied. For short term protection, oiling or greasing up a component may be adequate. However in the long term this is not satisfactory for the following reasons:

(a) the protective film will dry up and no longer seal the surface from the corrosive environment,

(b) oils and greases tend to absorb water from the atmosphere and corrosion can take place under the oil or grease,

(c) cheap oils and greases often contain active sulphur and acids which will themselves attack and corrode the very metal surfaces they are supposed to be protecting.

Therefore more sure and permanent protective coatings are required and some of the more common will now be considered briefly. For a more detailed study of the causes of corrosion and its prevention, see; *Engineering Materials, Vol. 2.*

Hot dipping

The finished components are cleaned, fluxed and immersed in molten metal (usually zinc). This is the principle of 'hot-dip galvanising'. It has the advantage over electrolytic galvanising of providing a thicker coating and, since it is applied to the finished work, it seals all the raw edges. In the case of agricultural hardware such as buckets, feeding troughs, etc., it has the added advantage of sealing the joints and making them water tight.

Electro-plating

This process is used to coat components with a thin layer of metal which may be protective, decorative or both. The components to be plated are immersed in an electrolyte and are made the *cathode* of the cell. That is they are connected to the negative terminal of a low voltage, heavy current, d.c. supply as shown in Figure 13.16. To complete the circuit *anodes* are immersed in the electrolyte and connected to the positive terminal of the supply. Upon the passage of an electric current metal ions are transferred from the electrolyte onto the surface of the cathode (work) to build up a protective metallic coating. The strength of the electrolyte is usually maintained by dissolving metal ions from the anode. The greater

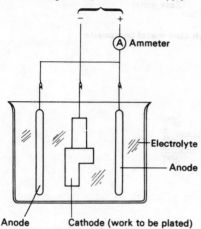

Low voltage heavy current d.c. supply

Ⓐ Ammeter

Electrolyte

Anode

Anode   Cathode (work to be plated)

Fig. 13.16   Electro-plating

the current and the longer it flows, the greater will be the mass of metal electro-plated onto the work.

Cladding

Originally this process was developed for the production of 'Sheffield Plate'. Prior to the development of electro-plating a thick coating of silver was applied to a sheet copper base by pressure welding a 'sandwich' of the metals between rolls. Nowadays it is used for the production of such metal composites as 'Alclad' where a corrosion resistant pure aluminium facing is pressure welded to a high strength aluminium base as shown in Figure 13.17.

Spraying

Corrosion resistant or wear resistant metals are melted in an oxy-acetylene flame or by electric arc and sprayed onto the component surface by compressed air blast. This process is generally used to build up worn bearing journals on shafts, although it is sometimes used on new components where a surface with special properties is required. The surface is rough and uneven and has to be finished by grinding.

Cementation

There are three processes in common use

(i) *Sherardising*. Cleaned and pickled plain carbon steel components are placed in a rotating steel barrel along with zinc metal powder and heated to 370°C for up to twelve hours. A zinc film is formed on the surface of the components which forms an excellent 'key' for subsequent painting.

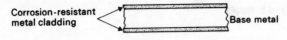

Corrosion-resistant metal cladding ← → Base metal

(a) Section through clad metal composite

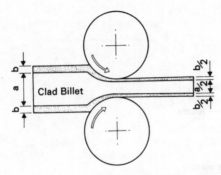

Clad Billet

(b) Proportional reduction of clad billet

Fig. 13.17 Cladding

(ii) *Calorising.* As above, except that aluminium powder is used at a higher temperature. This provides protection against high temperature oxidation rather than against ambient temperature corrosion, (e.g. car exhaust systems).

(iii) *Chromising.* (Not to be confused with 'chromating' — see below.) The work is baked in a mixture of aluminium oxide and metallic chromium powder at temperatures in excess of 1200°C in a hydrogen atmosphere to prevent oxidation. This is a very expensive process giving extreme protection against corrosion. It is used for chemical plant.

Anodising

Aluminium and aluminium alloy components are cleaned and degreased, after which they are etched, wire-brushed, or polished depending upon the surface texture required. The work is then made the *anode* of an electrolytic cell (see: electro-plating where the work is the *cathode*) and a direct current is passed through the cell. The electrolyte is a dilute acid and varies with the finish and protection required. Colours may be integral or applied subsequently by dyeing. The purpose of the treatment is to increase the thickness of the natural, protective oxide film and improve the corrosion resistance of the metal.

Chromating

(Not to be confused with chromising — see above.) This process is applied to magnesium alloy components which are dipped into a hot solution of potassium dichromate to form a hard, protective oxide film. The surface is then sealed by the use of zinc chromate paint, after which it is finished with a decorative paint.

Phosphating

Traditionally, phosphating processes were known by such names as: *Parkerising, Bonderising, Granodising*, and *Walterising*. More recently the whole range of phosphating processes have been standardised under BSS 3189. The components to be treated are pickled and degreased so that they are chemically clean and immersed in the hot, phosphating solution. The metal surface is converted into complex metal phosphates, chromates or oxides depending upon the process used. Finally they are sealed by oiling, waxing, or painting. The process is widely used as a pre-treatment for painting as it provides a good 'key' for the paint and prevents corrosion occurring under the paint film. Since phosphates have a high lubricity, phosphate conversion is often applied to bearing surfaces for gears, tappets, pistons, etc., and as a pre-treatment for metals which are subjected to severe cold-drawing processes.

Plastic coating

Plastic coatings can be functional, corrosion resistant and decorative. It provides the designer with a means of achieving: corrosion resistance; cushion coating (up to 6 mm thick); electrical and thermal insulation; flexibility over a wide range of temperatures; non-stick properties; permanent protection against weather and atmospheric pollution; reduction in maintenance costs; resistance to corrosion by a wide range of chemicals; the sealing of welds and porous castings, and decorative finishes.

The method of application depends upon the type of plastic to be used, its form of supply and the type of component to be coated. It is essential, in all cases, that the surface of the work piece is free from grease, dirt, oxidation or any other form of contamination.

Painting

A paint system consists of:

(i) A *primer* which is designed to 'key' firmly to the component being treated and which usually contains a corrosion inhibitor to prevent corrosion taking place under the paint film causing it to blister and flake off. As stated above, galvanising, sherardising, or phosphating is often used as a pre-treatment to provide an improved 'keying surface' for the primer and to prevent corrosion taking place under the paint film.

(ii) An *under-coat* to build up the paint film and give 'body' to the final colour. The under-coat must be built up to the required thickness by a number of thin coats rather than one thick coat. It is finally 'flatted down' to provide a suitably smooth surface for the finishing coat.

(iii) A *gloss-coat*. This is the finishing coat. It provides a hard, tough protective film which has a high gloss and the final colour. It seals the previous films against moisture and pollutants and is the most important element in the system for protecting the surface of the work.

Paint may be applied by brushing, spraying and dipping. It may be dried naturally or by stoving. The paint system must be chosen to suit the application process as well as the degree of protection required and the material and surface to be treated. Like all finishing processes, it is a highly complex technology if satisfactory protection is to be provided at a reasonable cost. Also, like all finishing processes, success depends upon the preparation of the surface to be treated to ensure it is clean, free from corrosion, and has a suitable texture to receive the finish. These finishes and their application will be considered in greater detail in *Engineering Materials, vol 2*.

13.14 Plastic degradation

Polymers degrade when exposed to heat, sunlight and weathering. They are also susceptible to 'solvent cracking' when exposed to the fumes of such substances as toluene, acetone and benzene (see: Section 13.5).

Degradation is usually accompanied by colour change, deterioration in mechanical properties, cracking and surface crazing. These changes usually occur as the result of chain division and cross-linking. Polymers based on olefins are particularly susceptible to such degradation. Overheating during the moulding process can also cause degradation.

Stabilisers can be added to reduce the rate of degradation and these include:

Antioxidants to prevent atmospheric oxidation:

Antiozonants to prevent ozone attack;

Ultraviolet filters to absorb the ultraviolet radiation of sunlight. The ultraviolet radiation of sunlight has sufficient energy to break down most of the chemical bonds in polymers. To prevent this, 'carbon black' is a pigment used as a radiation absorber in polythene film and sheet, but unfortunately it absorbs visible light as well as ultraviolet light. Where the black colour is unacceptable, Benzophenones and Benzotriazoles are used as ultraviolet filters. Although they are much more expensive, their spectrum transmission curves, which are shown in Fig. 13.18(*a*), compare very favourably with the ideal curve shown in Fig. 13.18(*b*).

The weathering properties of polymers must be carefully considered when selecting a material for outdoor use. Materials which give excellent results indoors may be quite unacceptable out of doors. For example, nylon is highly water absorbent and this reduces its suitability for use out of doors. A material which has a low cost may require more frequent replacement than one only a little more expensive. Although the former may be quite acceptable for indoor purposes, for outdoor applications it may be more economical in the long term to use the rather more expensive but more weather-resistant and longer-lasting material.

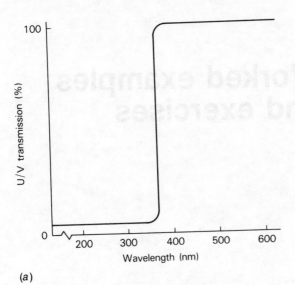

(a)

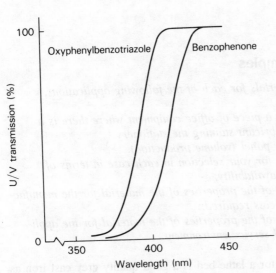

(b)

Fig. 13.18 Ultraviolet light absorbtion additives (a) Transmission curve for an ideal ultraviolet light absorbtion additive (b) Transmission curves for actual ultraviolet light absorbtion additives

14 Worked examples and exercises

14.1 Worked examples

1. *Select suitable materials for each of the following applications.*
    (a) Lathe bed.
    (b) Gear wheel for a piece of office equipment where there is a possibility of lubricant staining the stationery.
    (c) Motor car body panel (volume production).
    Discuss the reasons for your selection in each case in terms of:
    *   *(i) Cost and availability;*
    *   *(ii) Suitability of the properties of the material for the manufacturing process required;*
    *   *(iii) Suitability of the properties of the material for the application and service environment.*

(a) A suitable material for a lathe bed is a good quality grey cast iron as listed in BS 1452. Such a material is readily available at relatively low cost. It is suitable for casting to the intricate shape of a lathe bed in an ordinary sand mould. Although the casting pattern is an additional cost, this cost can be shared over all the castings produced. Grey cast iron is readily machined, although carbide tipped tools should be used as the surface scale is highly abrasive.

Cast iron is a stable material with a high rigidity so there is minimum deflection and loss of alignment in service. However it is advisable to 'weather' the casting naturally, or by heat treatment, after rough machining to remove residual stresses from the casting process and thus improve its dimensional stability. It has a low coefficient of friction so that the carriage and tailstock can easily slide along the bed-ways. Cast iron has inherent vibration damping properties which helps to improve the surface finish of the workpiece.

(b) A suitable material for the gear wheel for a piece of office material where lubrication is a problem would be Nylon. This material tends to absorb moisture from the atmosphere and this provides all the lubrication required. Nylon is widely available in a range of standard compositions which can be selected for specific purposes. It is relatively inexpensive and readily moulded to shape. Since machining is not required production costs are low providing that the number of gears required justifies the cost of the moulds. Its strength would be adequate for this application.

(c) Motor car body panels are pressed from low-carbon steel sheet. Such sheet is widely available because of worldwide demand in the volume car industry. Because the sheet is produced in large quantities the unit costs per sheet are low and this is a relatively low cost material. Low-carbon steel sheet in the bright annealed condition is easily pressed into complex shapes at high rates of production. It has a good surface finish suitable for painting to the standards expected on modern production cars.

2. *Briefly explain the difference between a crystalline solid and an amorphous solid and give an example of each type.*

Crystalline materials have their basic particles (atoms and molecules) arranged in strict geometric patterns called unit cells. These unit cells repeat the geometric pattern over and over again to build up the crystals of the material. Metals, in the solid state, are always crystalline. Amorphous materials have their basic particles arranged in a random manner and, because of this, do not exhibit a distinct melting temperature but soften gradually as their temperature is raised until they become liquid. Glass is typical of the amorphous solids in structure and behaviour.

3. *State the main difference between thermoplastic and thermosetting polymeric materials.*

Thermoplastic polymeric materials can be softened by heating both during the moulding process and subsequently and this limits their suitability for high temperature applications. There is no limit to the number of times thermoplastic materials may be softened by heating, although some degradation of the properties occurs if reheating occurs too frequently and at too high a temperature.

Thermosetting polymeric materials soften upon heating when initially moulded. During this high temperature moulding process a chemical change takes place in the material and it can never again be softened by reheating. This chemical change is called 'curing'.

4. *Describe the difference between:*

358

(a) a pure metal and an 'alloy';
(b) a substitutional solid solution and an interstitial solid solution.

(a) Pure metals consist of one type of atom only, for example: copper or aluminium. In practice, although very high levels of purity are obtained some residual impurities are always present and these tend to affect the properties of the metal.

'Alloys' are an intimate association of two of more component materials which form a metallic liquid or solid. The component materials may be metal elements or they may be metals and non-metals. The non-metals, in turn, may be pure substances or chemical compounds. Alloys may be 'tailored' to suit specific material requirements.

(b) In a substitutional solid solution one or more of the atoms in each unit cell of the crystals of the parent metal (solvent) is replaced by the alloying element (solute). It is from this substitution that this type of solid solution gets its name; for example, nickel atoms replacing some of the copper atoms in the face-centred cubic crystals of cupro-nickel alloys. In interstitial solid solutions the solute atoms are so small that they can lie in the spaces (interstices) between the solvent atoms. For example, carbon atoms can form an interstitial solid solution with face-centred cubic crystals of iron.

5. *With reference to Table 4.4, select a suitable steel for a low cost component with a hardness number of 210 H_B and a yield stress after heat treatment (R_E) of 420 MPa.*

Since the component is to be produced at low cost, a plain carbon steel would most likely be used. This would be relatively easy to form and machine. Reference to Table 4.4 (p. 100) shows that for a hardness condition of 210 H_B after heat treatment the steel will need to be in condition 'Q' or condition 'R'. In condition 'Q' the choice would be limited to 120M19, 150M19, and 216M28 by the yield stress requirement of 420 MPa. However in condition 'R' a much larger range of steels would be compatible with the requirements stipulated.

If the additional requirements of a limited ruling section of 60 mm and an Izod impact value of 50 J are added to the material specification, then the choice is restricted to 150M19. Since the question stipulated a 'low-cost' component, there would be no commercial justification for 'over-engineering' the component by stipulating a higher cost alloy steel with more sophisticated properties.

6. *With reference to the quench hardening of plain carbon steels, state what is meant by:*

(a) *the 'critical cooling rate';*
(b) *'mass effect'.*

(a) The critical cooling rate is the rate of cooling — or quenching —

which must be achieved or exceeded for hardening to take place. That is, the minimum rate of cooling at which the equilibrium transformations can no longer take place and at which the steel achieves a martensitic structure.

(b) In a thick component heat will be trapped at the centre of the material so that the critical cooling rate cannot be achieved throughout the material. This will result in the material being less hard at its centre than at its surface. This is referred to as 'mass effect'. The maximum diameter of a bar in which the critical cooling rate can be achieved throughout and in which the hardness is uniform throughout after quench hardening is referred to as the 'ruling section'.

7. *State the essential difference between a cast iron and a plain carbon steel.*

In a plain carbon steel all the carbon is chemically combined with the iron as iron carbide (cementite), or it forms a weak solid solution with iron (ferrite). The upper limit of carbon which can be so combined is theoretically 1.7 per cent. In practice it is limited to approximately 1.2 to 1.5 per cent to ensure all the carbon is combined under all conditions. In a cast iron the carbon content is substantially more than 1.7 per cent and the surplus carbon precipitates out on cooling and solidification to form flakes of graphite (an allotrope of carbon). These flakes are visible under the microscope and it is the discontinuities formed by the flakes which gives cast iron its mechanical properties. It is also this free carbon which gives cast iron its anti-friction properties.

8. *Select a suitable non-ferrous metal or alloy for each of the following applications giving reasons for your choice:*

    (a)   an electrical conductor which can be soldered;
    (b)   a die-cast car door handle;
    (c)   an internal combustion engine piston;
    (d)   a heavy duty bearing bush.

(a) High purity copper and aluminium are both excellent electrical conductors. However since ease of soldering is a specified criterion, only copper is applicable.

(b) Both aluminium and zinc alloys can be die-cast. In this example a zinc alloy such as 'mazak' would be most suitable. It casts at a relatively low temperature with less die wear than aluminium, it takes a sharp impression, it can be chrome plated to a high finish, and it has adequate strength for this application.

(c) A heat resistant aluminium alloy would be most suitable for this application. Such an alloy is BS 1490:LM14 (see Table 7.6). This is a good general purpose casting alloy suitable for pistons and cylinder heads

360

for liquid and air cooled internal combustion engines. It is relatively easy to die-cast and can be machined to a high finish and accuracy.

(d) White or bronze bearings can support heavy bearing loads. However white bearings require to be thin and to be backed up by high strength shells. This renders them unsuitable for self-supporting bushes and the preferred material would be phosphor bronze which combines good anti-friction properties with high strength.

9. *State the essential difference between the hot-working and cold-working of metals.*

Metals are said to be hot-worked when they are flow formed above the temperature of recrystallisation. Thus the crystals reform as fast as they are deformed and there is no tendency to work-harden. Less force is required to form the metal than when cold-working.

Metals are said to be cold-worked when they are flow formed below the temperature of recrystallisation. The crystals remain deformed and the metal becomes work-hardened. The amount of deformation is limited compared with hot-working and, unless process annealing is resorted to from time to time, excessive working can lead to the component cracking.

10. *State briefly what is meant by the terms:*

    (a)   *dry corrosion;*
    (b)   *wet corrosion;*
    (c)   *galvanic corrosion.*

(a) This is the direct oxidation of a freshly cut metal surface on exposure to atmospheric oxygen.

(b) This is either the oxidation of metals in contact with air and water as in the rusting of iron, or the reaction of metals and their surface oxides with the dilute acids found in rain as a result of burning fossil fuels. For example the formation of the green 'patina' on copper-clad roofs.

(c) This occurs when two dissimilar metals come together in the presence of an electrolyte such as salt water or the dilute acids found in rain. An electrolytic cell is formed and the more electro-negative metal will be eaten away.

14.2 Exercises

(Questions 1 to 4 inclusive are based mainly upon Chapter 1.)

1. Select suitable materials for each of the following applications and discuss the reason for your selection in each case in terms of cost

and availability, suitability for the manufacturing processes required, and suitability.

   (a)   National electricity grid overhead cables.

   (b)   Telephone handset.

   (c)   Body casting for a steam valve.

   (d)   Filing cabinet.

2.   With the aid of diagrams explain the essential differences between tensile strength, shear strength, and impact strength (toughness).

3.   Describe the main differences in magnetic properties between 'hard' and 'soft' magnetic materials.

4.   List THREE factors which affect the mechanical properties of materials and, in each case, give an example of how the factor chosen affects the properties.

(Questions 5 to 17 are based mainly on Chapter 2.)

5.   Sketch a typical representation of an atom and describe the particles which are to be found in the nucleus and orbiting around the nucleus.

6.   Briefly describe the essential differences between an element, a compound and a mixture.

7.   Briefly explain what is meant by the terms:

   (a)   allotrope, and

   (b)   isotope.

8.   Explain in detail the mechanism of crystal formation and growth as a metal cools from the liquid state to the solid state.

9.   Explain with the aid of sketches:

   (a)   the difference between *unit cell* and *space lattice*;

   (b)   how dendritic growth occurs;

   (c)   the difference between the crystal structure of a metal and its irregularly shaped grains.

10. Explain what is meant by the 'latent heat of fusion' and why the temperature of the metal remains constant whilst fusion occurs.

11. Explain the following terms associated with the solidification of a molten metal:

   (a)   inter-dendritic porosity;

   (b)   drawing;

   (c)   segregations;

   (d)   undissolved inclusions;

   (e)   gas porosity.

12. Explain what is meant by the term 'misorientation' and how this can affect the 'creep' resistance of a metal.

13. Describe the differences between the following groups of substances and explain how they affect the polymeric materials made from them:

   (a)   paraffins;

   (b)   olefins;

   (c)   naphthenes:

   (d)   aromatics.

14. Explain what is meant by crystallinity in polymeric materials and how crystallinity affects the properties of a polymer.

15. With the aid of diagrams show what is meant by the following terms related to polymer chains and how they affect the properties of the polymer:
    (a) linear;
    (b) branching;
    (c) cross-linked.

16. With reference to polymeric materials explain what is meant by:
    (a) uniaxial orientation;
    (b) biaxial orientation;
    (c) memory effect.

17. With reference to polymeric materials explain what is meant by:
    (a) glass transition temperature (T_g);
    (b) melting temperature (T_m).

(Questions 18 to 25 inclusive are based mainly on Chapter 3.)

18. Discuss the main advantages and limitations of metal alloys compared with pure metals.

19. Describe what is meant by an 'intermetallic compound' and how its properties are likely to vary from those of a solid solution.

20. Sketch typical examples of the following phase equilibrium diagrams for binary alloys:
    (a) simple eutectic type;
    (b) solid solution type;
    (c) combination type.

21. Draw the phase equilibrium diagram for cadmium-bismuth alloys. With reference to the diagram:
    (a) state the eutectic composition and describe the structure of the alloy in the solid state;
    (b) explain in detail the cooling of an alloy of composition 90 per cent cadmium and 10 per cent bismuth from above the liquidus to below the solidus, and describe its structure in the solid state.

22. Drawing the phase equilibrium diagram for copper-nickel alloys. With reference to the diagram:
    (a) explain in detail the cooling of an alloy of composition 60 per cent copper and 40 per cent nickel from above the liquidus to below the solidus, and describe its structure in the solid state;
    (b) explain why this series of alloys do not show a eutectic composition.

23. Draw the phase equilibrium diagram for tin-lead alloys.
    (a) Indicate on the diagram: (i) the liquidus; (ii) the solidus; (iii) the solvus.
    (b) Explain in detail the cooling of an alloy of composition 60 per

cent lead and 40 per cent tin from above the liquidus to below the solidus and describe its structure in the solid state.

(c) Explain briefly why 'plumbers solder' has a high lead content, and why solder used for securing electronic components has a eutectic composition.

24. With reference to the copper-nickel phase equilibrium diagram, explain what is meant by 'coring' and what is meant by 'diffusion'.

25. With reference to the copper-aluminium phase equilibrium diagram, explain what occurs in the alloy during 'solution treatment' and 'precipitation hardening'.

(Questions 26 to 33 inclusive are based mainly on Chapter 4.)

26. State briefly what is meant by the terms:
    (a) ferrite;
    (b) pearlite;
    (c) cementite;
    (d) austenite.

27. A typical plain carbon steel is specified as BS970:080M40. From this specification derive the composition of the steel.

28. If the heat treatment condition 'Q' is added to the specification in question 27, list the ruling section and the properties which could be expected of this steel.

29. Show by means of a diagram the effect of carbon content on the hardness, tensile strength and ductility of annealed plain carbon steels.

30. Sketch the 'steel section' of the iron-carbon phase equilibrium diagram for plain carbon steels and describe the cooling from the austenitic condition to room temperature for a:
    (a) hyper-eutectoid steel;
    (b) hypo-eutectoid steel;
    (c) steel of eutectoid composition.

31. Describe the effects of the following elements when included, by accident or by design, in plain carbon steels:
    (a) manganese;
    (b) silicon;
    (c) sulphur;
    (d) phosphorus.

32. With reference to the iron-carbon phase equilibrium diagram:
    (a) explain what is meant by the following change points: (i) A_{r1}, (ii) A_{r3}, (iii) A_{rcm};
    (b) explain the difference between A_r points and A_c points;
    (c) explain what is meant by the 'Curie point'.

33. Since iron is allotropic:
    (a) state the types of crystal it will form: (i) above 1400 °C; (ii) below 910 °C; (iii) between 910 °C and 1400 °C.

(b) explain what is meant by 'recalescence', and describe how this phenomenon may be demonstrated.

(Questions 34 to 43 inclusive are based mainly on Chapter 5.)

34. Briefly explain the following terms as applied to the heat treatment of plain carbon steels:
    (a) quench hardening;
    (b) tempering;
    (c) full annealing.
35. (a) Describe how a chisel made from 0.8 per cent carbon steel may be hardened and tempered.
    (b) How can the heat treatment temperatures be judged visually in this example?
    (c) Why should a long, slender component be quenched with its axis vertical?
36. With reference to the quench hardening of plain carbon steels, explain what is meant by the term 'critical cooling rate'.
37. With the aid of diagrams explain in detail what is meant by:
    (a) recrystallisation;
    (b) cold-working;
    (c) hot-working.
38. With reference to plain carbon steels, explain in detail the essential differences between:
    (a) stress relief (process) annealing;
    (b) spheroidising annealing;
    (c) full annealing:
    and give one example in each instance where the process would be used.
39. With reference to the steel section of the iron-carbon phase equilibrium diagram, explain how:
    (a) a hypo-eutectoid steel can be heat treated to give maximum toughness. Describe the micro-constituents present in the steel after heat treatment.
    (b) a hyper-eutectoid steel can be heat treated so that it is suitable for cutting metal. Describe the micro-constituents present in the steel after heat treatment.
40. With reference to the iron-carbon phase equilibrium diagram, explain how a low carbon steel can be case hardened by the pack-carburising process so that the components have a hardened and tempered case and a toughened, fine grain core.
41. With the aid of diagrams explain:
    (a) how selected areas (e.g. screw threads) of a case hardened component can be left soft for subsequent machining;
    (b) the principles of surface hardening by: (i) flame hardening; (ii) induction hardening, and give an example where each of these processes would be appropriate.

42. Explain fully what is meant by:
    (a) limited ruling section;
    (b) mass effect;
    (c) hardenability;
    as applied to plain carbon steels and how the 'Jominy end quench test' can be used to assess hardenability.
43. Discuss the advantages and limitations of the 'nitriding process' for applying a wear resistant surface to engineering components.

(Questions 44 to 50 inclusive are based mainly on Chapter 6.)

44. With reference to BS 1452:1977 list the following properties of a grade 220 grey cast iron:
    (a) tensile strength;
    (b) 0.1 per cent proof stress;
    (c) compressive strength;
    (d) shear strength.
    Why does this standard NOT give the composition for each grade of cast iron listed in it?
45. Briefly explain the effect of the following alloying elements on cast irons:
    (a) chromium;
    (b) copper;
    (c) vanadium.
46. Discuss in detail the following malleablising processes as applied to cast irons, together with typical properties (BS 6681:1986) and applications.
    (a) Whiteheart process.
    (b) Blackheart process.
    (c) Pearlitic processes.
47. Discuss the effects on the structure, and properties of the following heat treatment processes as applied to cast irons:
    (a) annealing;
    (b) stess relieving.
48. *(a)* Sketch the cast iron section of the iron-carbon phase equilibrium diagram and label the more important compositions and temperatures.
    (b) With reference to the above diagram, distinguish between the formation of ferritic grey cast iron and pearlitic grey cast iron.
49. Describe with the aid of diagrams the essential difference between grey cast iron and spheroidal graphite (SG) cast iron and compare their advantages and limitations.
50. State the compositions of a suitable cast iron for each of the following applications giving reasons for your choice.
    (a) A lightly stressed ornamental casting.
    (b) An internal combustion engine crankshaft.
    (c) A machine tool bed.

(d) A mains water pipe.
(e) Brake discs.
(f) A furnace flue damper.

(Questions 51 to 59 inclusive are based mainly on Chapter 7.)

51. State the main alloying elements in:
    (a) brass;
    (b) duralumin;
    (c) gun-metal;
    (d) phosphor bronze.

52. State an example of each of the following types of aluminium alloy and in each case, state the composition and a typical application:
    (a) non-heat-treatable wrought alloy;
    (b) non-heat-treatable casting alloy;
    (c) heat-treatable wrought alloy;
    (d) heat-treatable casting alloy.

53. State the effect on pure copper of adding the following alloying elements:
    (a) tellurium;
    (b) beryllium;
    (c) cadmium;
    (d) arsenic.

54. (a) Sketch the aluminium-silicon phase equilibrium diagram and explain in detail the changes which take place as a 6 per cent silicon alloy is cooled from the molten state to room temperature.
    (b) Discuss the need for 'modification' of aluminium-silicon alloys by the addition of metallic sodium, and show the effect of such modification on the aluminium-silicon phase equilibrium diagram.

55. Discuss in detail the differences in composition, properties and applications of:
    (a) high-conductivity copper;
    (b) fire refined tough pitch copper;
    (c) oxygen-free high-conductivity copper;
    (d) phosphorus deoxidised copper.

56. (a) State the composition, properties and typical applications of the aluminium alloy known as 'duralumin'.
    (b) Describe in detail, with reference to the aluminium-copper phase equilibrium diagram, the softening of duralumin by solution treatment, and the subsequent hardening of duralumin by natural aging (precipitation).

57. Sketch the copper-zinc phase equilibrium diagram and refer to it when describing the differences in composition, properties and typical applications of:
    (a) the α brasses;
    (b) the duplex brasses ($\alpha + \beta'$).

58. Sketch the copper-tin phase equilibrium diagram and refer to it when describing the differences in composition, properties and typical applications of:
    (a) the α phase bronze alloys;
    (b) the duplex bronze alloys (α + δ).

59. Describe the composition, properties and a typical application of:
    (a) a magnesium alloy;
    (b) a zinc alloy suitable for die-casting.

(Questions 60 to 68 inclusive are based mainly on Chapter 8.)

60. Select a suitable polymeric material for the following applications giving reasons for your choice:
    (a) flexible electric cable insulation;
    (b) non-stick anti-friction coating;
    (c) high-strength light-weight ropes for mountaineering;
    (d) low friction, oil-less bushes for food processing machinery;
    (e) heat insulating panels.

61. (a) Explain what is meant by an 'elastomer' and how it differs from other polymeric materials.
    (b) State typical applications for the following elastomers giving reasons for your choice:
        (i) styrene-butadiene-rubber (SBR);
        (ii) polyisoprene (natural) rubber;
        (iii) polychloroprene (neoprene) rubber;
        (iv) silicone rubber;
        (v) acrylic rubber.

62. State the reasons for the inclusion of the following additives in a thermosetting moulding powder:
    (a) filler;
    (b) pigment;
    (c) catalyst;
    (d) accelerator;
    (e) mould release agent.

63. State typical applications for the following filler materials:
    (a) glass fibre;
    (b) wood flour;
    (c) calcium carbonate;
    (d) aluminium powder;
    (e) shredded paper;
    (f) mica granules.

64. With the aid of a diagram explain how a typical thermosetting moulding powder is cured (polymerised) in the mould by the condensation of water molecules. Describe the precautions which have to be taken in the design of a mould to compensate for the curing process.

65. Contrast and compare the general properties of polymeric materials with those of the more common metals.

368

66. Explain why the following additives are used in association with polymeric materials:
    (a) plasticisers;
    (b) stabilisers;
    (c) antistatic agents;
    (d) antioxidants.
67. Compare the properties of the following thermoplastic materials and state a typical application for each:
    (a) polyethylene;
    (b) polystyrene;
    (c) unplasticised polyvinyl chloride;
    (d) polymethyl methacrylate;
    (e) cellulose acetate.
68. Compare the properties of the following thermosets and state a typical application for each:
    (a) phenol formaldehyde;
    (b) urea formaldehyde;
    (c) epoxy resin;
    (d) alkyd resin.

(Questions 69 to 77 inclusive are based mainly upon Chapter 9.)

69. With the aid of diagrams explain the difference between sliding and rolling bearings and give a typical example of each.
70. Discuss, in detail, the main causes of wear in a bearing.
71. State the essential properties of bearing materials:
    (a) for sliding bearings;
    (b) for rolling bearings.
72. Compare and contrast the properties and applications of 'white' bearing metals and 'tin-bronze' bearing metals.
73. Explain the importance of lubrication in bearings and discuss the essential properties of a lubricant for medium duty sliding bearings.
74. Compare and contrast the properties and uses of metallic and non-metallic bearing materials.
75. (a) Discuss the significance of surface contamination in bearing materials.
    (b) Discuss the combination of similar and of dissimilar materials in bearings.
76. Discuss the significance of surface coatings on bearing materials in terms of wear resistance and embedability.
77. Select a suitable material for each of the following bearings and indicate any surface treatment which may be an advantage:
    (a) a drill bush;
    (b) a main bearing shell for a car engine;
    (c) the balls and races of an anti-friction bearing;
    (d) a water pump impeller bearing;
    (e) the 'brasses' for a plumber block-type journal bearing.

(Questions 78 to 84 inclusive are based mainly on Chapter 10.)

78. Draw a section through a typical sand mould for a hollow casting and describe the essential parts of the mould and the technique for producing the mould.

79. Describe the following defects which may be found in castings:
    (a) blowholes;
    (b) porosity;
    (c) scabs;
    (d) fins;
    (e) cold shuts;
    (f) uneven wall thickness;
    (g) drawing.

80. Discuss the advantages and limitations of shaping material by:
    (a) machining from the solid;
    (b) forging to shape.

81. Describe THREE hot-working processes which exploit the property of malleability in a metal.

82. (a) Describe TWO cold-working processes which exploit the property of ductility in a metal.
    (b) Describe TWO cold-working processes which exploit the property of malleability in a metal.

83. With the aid of sketches explain the principles of injection moulding polymeric materials and list the criteria essential for the production of a successful moulding. State the group of polymeric materials which are usually injection moulded.

84. Draw a section through a simple mould suitable for thermosetting polymeric materials and explain the importance of the following features:
    (a) ejector;
    (b) vents;
    (c) flash gutter;
    (d) flash land;
    (e) draught.

(Questions 85 to 94 inclusive are based mainly on Chapters 11 and 12.)

85. With the aid of diagrams describe the difference between ultimate tensile stress (UTS) and proof stress (PS) when determining the properties of a ductile metal.

86. (a) Sketch a typical tensile test curve for an annealed low-carbon steel and indicate the following on the curve: (i) elastic range; (ii) elastic limit; (iii) yield point; (iv) plastic range; (v) breaking point of the specimen.
    (b) Describe how the ductility of a material can be calculated from a tensile test in terms of elongation percentage.

87. (a) With the aid of sketches explain how the *secant modulus* for polymeric materials is determined.

(b) Explain why *creep* is such an important factor when testing polymeric materials and distinguish between creep and instantaneous elongation.

88. *(a)* With the aid of sketches show the essential differences between the Charpy and Izod impact tests, paying particular attention to the proportions of the specimens, the notch profile, method of support, and criteria for determining the impact number.

(b) Describe how the results of an impact test can be interpreted to indicate the toughness of a material, its susceptibility to crack propagation, and how the appearance of the fractured surfaces differs between ductile and brittle materials.

89. With the aid of sketches describe:

(a) how the Brinell hardness test is performed;

(b) the relationship between load and the diameter of the indenter and how the hardness number is derived;

(c) any precautions which should be taken to ensure that the test result is valid;

(d) how properties other than hardness may be derived by calculation and visual inspection of the test results.

90. Compare and contrast the Brinell hardness test, the Vickers hardness test, and the Rockwell hardness test. Pay particular attention to the essential differences between the tests, suitable applications, and the dangers inherent in using hardness conversion tables to compare the results of these tests.

91. Explain why non-destructive testing is an essential element of a programme of quality control, and give examples of where it would be used.

92. Compare the advantages and limitations of the following inspection techniques for finding surface and internal defects in components:

(a) dye penetrants;

(b) ultrasonic testing;

(c) eddy-current testing;

(d) magnetic testing.

93. *(a)* Compare the advantages and limitations of gamma-ray and X-ray sources of radiation for the radiographic inspection of metal components and assemblies, explain where each would be used, and explain what is meant by *half life thickness*.

(b) Describe in detail the safety precautions which need to be taken when using radiography as an inspection technique.

94. Select a suitable test or group of tests, giving reasons for your choice, for detecting the following faults:

(a) a misplaced insert in a plastic moulding;

(b) slag inclusions in a casting;

(c) surface cracks in a forging;

(d) porosity in a pipe weld;

(e) inclusions and/or discontinuities orientated parallel to the axis of a bright drawn steel bar.

(Questions 95 to 100 inclusive are based mainly upon Chapter 13.)

95. With the aid of a diagram distinguish between primary, secondary and tertiary creep and discuss the importance of creep in highly stressed engineering components operating at elevated temperatures.

96. With the aid of sketches describe how a fatigue test may be performed and the appearance of a typical S/N curve derived from such a test. Explain how the S/N curve is interpreted.

97. *(a)* With the aid of sketches explain what is meant by the terms: fluctuating load; pulsating load, and alternating load when applied to fatigue testing.

   (b) Discuss FOUR factors which affect the fatigue resistance of a component.

98. *(a)* Describe what is meant by the terms:
   (i) atmospheric corrosion:
   (ii) galvanic corrosion.

   (b) Discuss FOUR factors which affect the corrosion of metallic components and assemblies.

99. *(a)* Describe in detail how metals such as aluminium, copper and zinc resist corrosion.

   (b) Discuss the main factors responsible for the degradation of polymeric materials in service.

100. *(a)* Explain how low-carbon steel components can be protected by the process of galvanising and explain why the coating is said to be 'sacrificial'.

   (b) Explain why the surface of a metal component must be carefully prepared prior to painting and how this may be performed.

   (c) Explain how paint systems are classified according to the method by which the paint dries and hardens, and describe the various coats which are used to build up a complete paint system.

14.3 Selection and application of materials

The use of the foregoing information in the selection and application of engineering materials for various components is an essential exercise for all engineers. Figure 14.1 shows an exploded view of a typical lathe tailstock, and the following exercises are based upon the selection of materials for various components from that drawing. Components requiring materials not included in this text will be disregarded. In all the following exercises give reasons for your choice of materials.

1. Select a suitable material for component 77927-0 (casting). What precautions must be taken:

   (a) to ensure uniform cooling and minimum internal stresses;
   (b) to ensure dimensional stability in service?

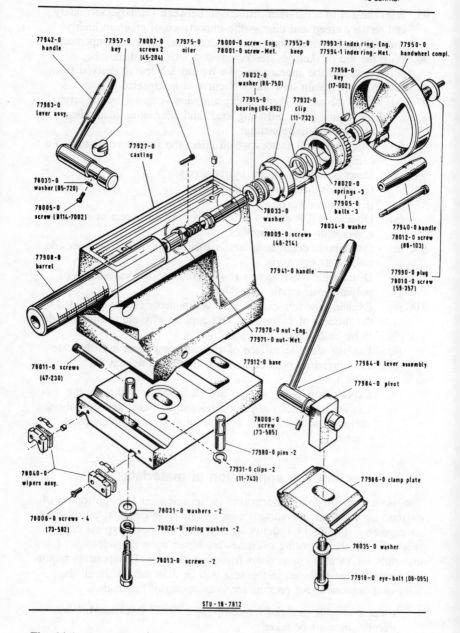

372

77942-0 handle

77957-0 key

78007-0 screws 2 (45-204)

77975-0 oiler

78000-0 screw - Eng. 78001-0 screw - Met.

77953-0 keep

77993-1 index ring - Eng. 77994-1 index ring - Met.

77950-0 handwheel compl.

78032-0 washer (86-750)

77958-0 key (17-002)

77963-0 lever assy.

77915-0 bearing (04-892)

77932-0 clip (11-732)

77927-0 casting

78030-0 washer (85-720)

78005-0 screw (8114-7002)

77900-0 barrel

78020-0 springs - 3

77905-0 balls - 3

78033-0 washer

78009-0 screws (46-214)

78034-0 washer

77940-0 handle

78012-0 screw (88-103)

77941-0 handle

77990-0 plug 78010-0 screw (59-357)

77970-0 nut - Eng. 77971-0 nut - Met.

78011-0 screws (47-230)

77912-0 base

77964-0 lever assembly

77984-0 pivot

78008-0 screw (73-585)

78040-0 wipers assy.

77980-0 pins - 2

77931-0 clips - 2 (11-743)

78006-0 screws - 4 (73-582)

78031-0 washers - 2

78026-0 spring washers - 2

78013-0 screws - 2

77986-0 clamp plate

78035-0 washer

77918-0 eye-bolt (08-095)

STU - 10 - 7812

Fig. 14.1 Tailstock

2. Select a suitable material for component 77942-0 (handle).
3. Select a suitable material for component 77970-0 (nut), paying particular attention to its anti-friction properties and the relative wear between the screw and the nut.
4. Select a suitable material for component 77908-0 (barrel) and describe any heat treatment which may be necessary to ensure toughness and wear resistance.
5. Select a suitable material for component 77993-1 (index ring) and describe any surface finish treatment which may be required to ensure that the engraved scale is easy to read and wear resistant.
6. Select a suitable material for component 77950-0 (handwheel).
7. Select suitable materials for 78040-0 (wiper assembly).
8. Select a suitable material for component 77953-0 (keep) paying particular attention to cost and ease of production since this component is not subject to wear. From the drawing this component would appear to have its bore bushed with an antifriction metal to support component 78000. Specify a suitable bearing metal for this bush.
9. Select a suitable material for the component 77986-0 (clamp plate). Note: this component is subject to bending forces.
10. Select a suitable material for component 78000 (screw) and specify any heat treatment which may be necessary. Pay particular attention to any problems of cutting the screw, wear, and cost of replacement relative to the nut with which it mates.

Index

378